# Essentials of Natural Gas Microturbines

# Essentials of Natural Gas Microturbines

Valentin A. Boicea

CRC Press is an imprint of the
Taylor & Francis Group, an **informa** business

This book contains republished or adapted works of the U.S. Government in Chapters 1, 2, 4, 5, and 8.

**Front cover:** The Capstone C200 cutaway. (Courtesy: Capstone Turbine Corporation®, Chatsworth, California.)

CRC Press
Taylor & Francis Group
6000 Broken Sound Parkway NW, Suite 300
Boca Raton, FL 33487-2742

First issued in paperback 2017



Version Date: 20131025

ISBN 13: 978-1-4665-9471-5 (hbk)
ISBN 13: 978-1-138-07418-7 (pbk)



**Library of Congress Cataloging-in-Publication Data**

Boicea, Valentin A.
Essentials of natural gas microturbines / Valentin A. Boicea.
pages cm
Includes bibliographical references and index.
ISBN 978-1-4665-9471-5 (hardcover : alk. paper)
1. Gas-turbine power-plants. I. Title.

TK1076.B65 2014
621.43'3--dc23 2013038210

**Visit the Taylor & Francis Web site at**
**http://www.taylorandfrancis.com**

**and the CRC Press Web site at**
**http://www.crcpress.com**

*PARENTIBVS OPTIMIS IOHANNI ET AEMILIAE BOICEIIS,*

*SCIENTIAEQVE PERITISSIMIS IOHANNI FRANCISCO*

*CHICCO ET PHILIPPO SPERTINO, CETERISQVE*

*AMICIS ET COLLEGIS ITALICIS GRATO ANIMO*

**ADINAE AMATISSIMAE**

*Domnului Antonel…*

# Contents

# Acknowledgments

The author wishes to thank Prof. Dr. Eng. Mihaela Marilena Albu, Faculty of Electrical Engineering, University Politehnica of Bucharest; Prof. Dr. Eng. Ecaterina Andronescu, chair of the University Politehnica of Bucharest Senate; Prof. Dr. Eng. Gabriel Bazacliu, Faculty of Power Engineering, University Politehnica of Bucharest; Nancy A. Blair-DeLeon, senior manager, Author Engagement and Content Discoverability at IEEE; Prof. Michael Hewson Crawford, University College, London; Kathryn Everett, project coordinator for Taylor & Francis; Adam Gottlieb, assistant executive director, California Energy Commission; Anne Hampson, senior associate at ICF International; Prof. Dr. Eng. Ştefan Kilyeni, Faculty of Electrical and Power Engineering, University Politehnica Timişoara; Rob Oglesby, executive director, California Energy Commission; Fernando Pina, office manager, California Energy Commission; Jonathan Plant, executive editor in Engineering, Taylor & Francis; Prof. Dr. Eng. Claudia Laurenta Popescu, vice chancellor of the University Politehnica of Bucharest; Maria A. Silva, marketing assistant, Capstone Turbine Corporation; and E. Harry Vidas, vice president, ICF International.

# 1

# Gas Turbines and the Automotive Industry

Microturbines (MTs) running on natural gas represent an important and emerging technology in distributed generation (DG) systems. Natural gas MTs can be an appealing choice for the customer given their relatively high efficiency (approximately 33% or even 80% in some cases when they are used in combined heat and power [CHP] applications) compared to other types of DG equipment. Another important advantage of these units is the fact that they can also be used as a backup resource for other DG systems such as wind farms. When the wind speed drops to approximately below 6 m/s, natural gas MTs can be used to avoid blackouts and improve the grid stability.

As will be discussed in the next chapters, this kind of equipment is based on three key components: the engine, which can be driven either on liquid fuel or gas; the fuel system, comprising the gas boost compressors (which feed the MT with gas at the appropriate pressure), and, finally the power converter, providing electrical energy at 50 Hz or 60 Hz depending on the application (see Chapter 5).

From a historical point of view, gas turbines having a power range between 25 and 400 kW were first used during the 1950s and 1960s as car engines by manufacturers such as Rover®, Leyland Motor Corporation Limited® (a subsidiary of Rover), FIAT®, Ford®, Chrysler®, GM® (General Motors), and Austin®. Unfortunately, all of these engines used very expensive reduction gear boxes, making them economically unfeasible, especially when compared to the reciprocating engines. Despite this, the technology flourished in the aerospace industry, being used for the fabrication of auxiliary power units (APUs) which start the main engines of the airplane and operate its electrical systems (like, for instance, the accessories when the main engines are shut down). This is why gas turbines are currently considered aero-derivative. The major disadvantage of this is the low number of units produced per year (approximately 1,000) due to the high production costs resulting from quality requirements in this field. Moreover, the life of these engines is limited to 10,000 hours, making them unsuitable for cogeneration or prime power applications [Moo02]. The situation has changed in the last decade, and the technology of gas turbines (especially the power electronics and the design of the machine itself) has been much improved, making this kind of equipment more viable and more accessible to end users [Moo02].

To represent an attractive alternative to other DG systems, MTs have to be robust, competitive from a cost point of view, and have acceptable efficiency. The first units having the same rated power as the ones used today in DG came into use in the automotive industry. Like the units used for DG, these gas turbine engines provided little vibration, efficient torque curves, low maintenance costs, and good fuel adaptability. As observed in the following pages, they could be operated with a wide range of fuels like fuel oil, paraffin, diesel oil, kerosene, or even unleaded gasoline.

Adoption of this equipment for car engines was limited by:

- Slow response of the fuel regulator valve
- Elevated fuel consumption at partial load
- Elevated fuel consumption at full load

A general flow diagram of a gas turbine for powering a vehicle is presented in Figure 1.1. As shown in the figure, the air enters the compressor and is then preheated in the recuperator. In the next stage, the preheated air is mixed with fuel in the combustor, and the resulting burning products will flow first through the gas generator turbine and then through the power turbine, from which they are finally dispersed into the atmosphere through the recuperator. The power turbine is attached through a power shaft to a reduction gear box which is further connected through an output shaft to a suitable load, like for instance the wheels of a vehicle (not illustrated here).

## 1.1 The Fuel Control System for a Gas Turbine Engine Developed by Rover

The Rover fuel control system for a gas turbine engine, the first ever developed for a vehicle, is very straightforward. Its core consists of the fuel pump and a valve which are engine driven [DBr46][RLa48][RNP63]. This valve

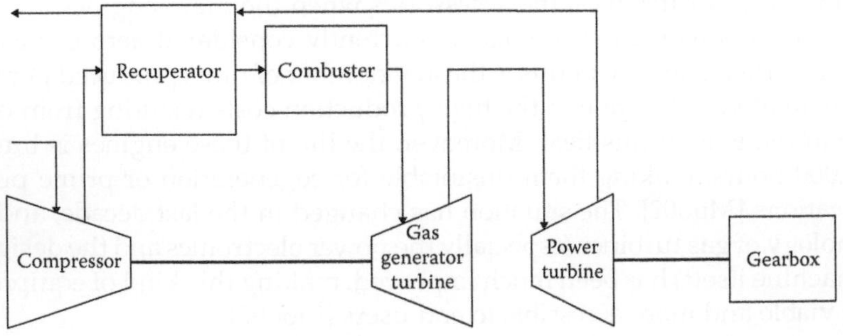


**FIGURE 1.1**
General flow diagram of a gas microturbine (MT) used in the automotive industry. (Courtesy of Robin Mackay, Agile Turbine Technology, Manhattan Beach, CA.)

is centrifugally operated and is capable of bypassing fuel from the engine burner back to the pump inlet at a certain engine speed [RLa48][RNP63].

In principle, the pump is rotary and highly efficient. Another important feature of this equipment is a thermal-operated valve that opens and lets the fuel pass when a certain area of the motor reaches a predetermined temperature [CSK59][RLa48][RNP63]. Theoretically, this type of valve is optional.

Another key component of the fuel control system is the pressure-operated valve. This is in fact a poppet valve that is pushed by a spring toward its closed position and programmed to open at a value of the fuel pressure greater than the spring loading. The spring loading, on the other hand, does not have a fixed value. This can be modified by means of a screw, mounted at the end of the spring [CSK59][DBr46][RNP63]. This kind of setup permits control of the fuel flow to the burner at a certain pump speed interval, which is below the speed of the centrifugally operated valve [CSK59][RNP63]. The engine has one or more fuel atomizers which will inject in this way an optimum fuel flow into the combustion chamber.

As shown in Figure 1.2, the pump rotor has three bores, each containing a piston that during a single revolution of the rotor will suck in fuel from the car reservoir, and at a later stage will drive it out into the fuel outlet

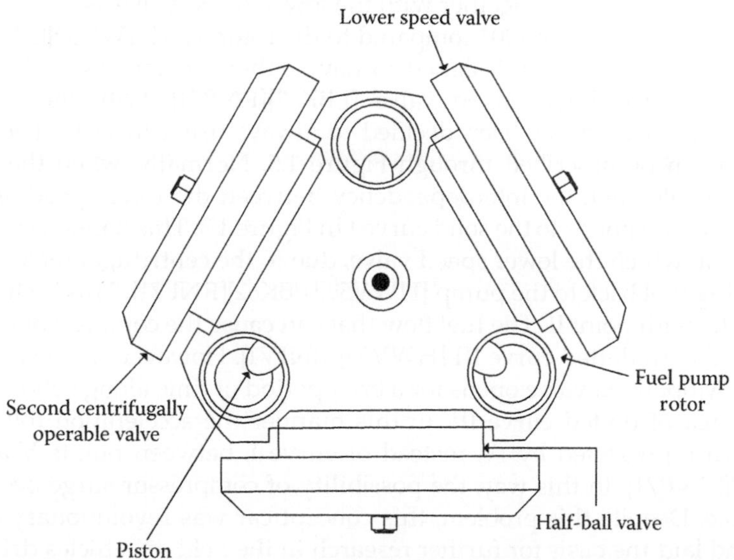


**FIGURE 1.2**
Approximate representation of the Rover fuel pump rotor for a gas turbine engine. (From P.B. Kahn, Fuel control governors. U.S. Patent 3,035,592, issued May 22, 1962, available online: http://www.uspto.gov [accessed March 5, 2013]; R.N. Penny, Gas turbine engine fuel system. U.S. Patent 3,085,619, issued April 16, 1963, available online: http://www.uspto.gov [accessed March 4, 2013]; and R.N. Penny, Rotary fuel pump. U.S. Patent 3,594,100, issued July 2, 1971, available online: http://www.uspto.gov [accessed March 4, 2013.] With permission.)

[CSK59][DBr46][RNP63]. These pistons are driven by an inclined, fixed cam plate [RNP63].

Another characteristic of the Rover fuel control system is that it can include a second centrifugally operable valve (different from the one already mentioned and connected in parallel with this one), initially having the role of keeping the engine speed within a certain range [DBr46][RNP63]. This second centrifugally operable valve will be fully investigated below. The most important characteristic of this system is that the fuel pump is connected to the shaft of the car motor, and in this way the fuel pump speed becomes a function of the vehicle motor speed [HWV59][RNP63][RNP71].

Unfortunately, after the first tests, it has been observed that when the motor and the fuel pump were accelerated before the opening of the valve which bypassed back the fuel to the pump, there was a tendency for the compressor to surge [PBK62][RNP71]. To avoid such a situation, the use of the second centrifugally operated valve became imperative. Thus, concurrently with maintaining the engine speed within a certain range, this second valve had the role of opening temporarily, hence reducing the fuel flow to the engine.

In Figure 1.2, one can distinguish, besides the fuel pump rotor, a half-ball valve attached to a leaf spring (not shown), mounted parallel to the rotor axis and connected to this one with screws. This half-ball valve is actuated at high rotor speed. This valve, together with the lower speed valve, is connected to the rotor at a position of 120° compared to the rotor axis [HWV59][RNP71].

Initially, the rotor was designed to have either a balancing weight or a second centrifugally operated valve [PBK62][RNP71]. With the problem mentioned above, it has been decided to always use a throttle. The surge problem can be described through Figure 1.3. Normally, when the pump rotor is accelerated, the interdependency between the rotor speed and the fuel flow corresponds to the solid curve I in Figure 1.3. This happens until the moment at which the lower speed valve, due to the centrifugal force, opens, bypassing fuel back to the pump [HWV59][PBK62][RNP71]. This is shown in Figure 1.3 with point II. The fuel flow that can cause the compressor surge is indicated with dotted curve III [HWV59][RNP71]. However, the second centrifugally operated valve opens for a brief period of time along solid curve I, in the area of dotted curve III. In this manner the acceleration follows in a pattern represented by IV, instead of curve I, between points V and VI [PBK62][RNP71]. In this way, the possibility of compressor surge occurring is remote. Despite this problem, this concept car was revolutionary at that time and laid the basis for further research in the field of vehicles driven by gas turbine engines.

For more detailed information regarding the fuel control system for a gas turbine engine implemented by Rover, please refer to [CSK59][DBr46][HWV59][PBK62][RLa48][RNP63][RNP71].

The car that benefited from such a system was capable of achieving 97 km/h in just 14 seconds (which was a performance at that time) but unfortunately consumed 1 L of fuel at every 2 or 2.5 km [BBC-- ].

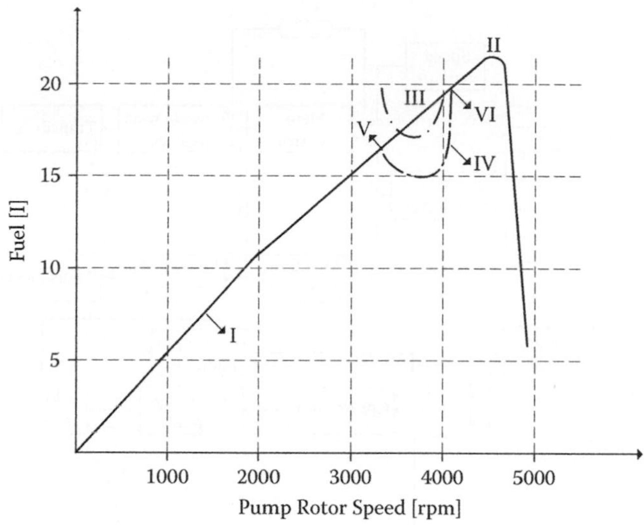


**FIGURE 1.3**
Approximate view of the relation between the fuel flow and the fuel pump rotor speed. (From H.W. Van Gerpen, Hydraulic apparatus. U.S. Patent 2,892,311, issued June 30, 1959, available online: http://www.uspto.gov [accessed March 5, 2013]; P.B. Kahn, Fuel control governors. U.S. Patent 3,035,592, issued May 22, 1962, available online: http://www.uspto.gov [accessed March 5, 2013]; and R.N. Penny, Rotary fuel pump. U.S. Patent 3,594,100, issued July 2, 1971, available online: http://www.uspto.gov [accessed March 4, 2013]. With permission.)

The fuel used was paraffin, petrol, or diesel oil [BBC-- ]. The Leyland truck, on the other hand, developed at the end of the 1960s, used an improved system compared to the one described above, since the company that produced it was a subsidiary of Rover.

## 1.2 The Fuel Control System for a Gas Turbine Engine Developed by FIAT

The control system for a gas turbine engine implemented by FIAT (Fabbrica Italiana Automobili Torino—the Italian manufacturer of automobiles from Turin) is also among the first ever made and has a straightforward principle.

It consists mainly of the fuel flow control to the combustion chamber. This is accomplished through the speed control of the gas turbine as well as through temperature control of the exhaust gases during the whole operating period of the motor [ALo70][GCa76][RBo61].

This gas turbine control takes place based on electric signals proportional to different parameters like temperature, pressure, or speed which are

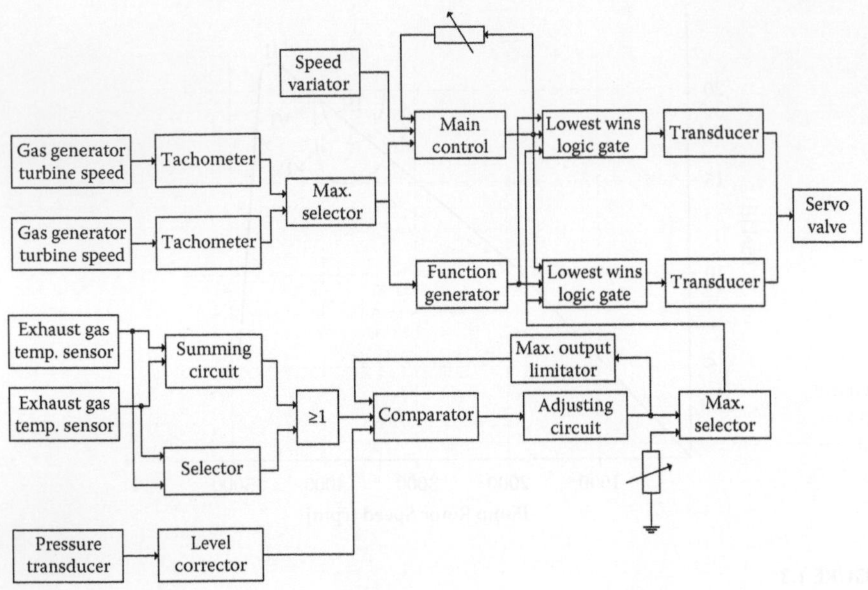


**FIGURE 1.4**
Approximate view of the electronic device controlling the fuel system for the gas turbine engine developed by FIAT. (From G. Caire et al., Device for controlling gas turbine engines. U.S. Patent 3,987,620, issued October 26, 1976, available online: http://www.uspto.gov [accessed February 27, 2013]; R. Bodemuller et al., Gas turbine acceleration control. U.S. Patent 2,971,338 issued February 14, 1961, available online: http://www.uspto.gov [accessed February 28, 2013]; and W. Rowen et al., Constant power control system for gas turbine. U.S. Patent 3,639,076, issued July 14, 1972, available online: http://www.uspto.gov [accessed February 28, 2013]. With permission.)

equivalent to the operating conditions of the engine. These signals are then directed to a processor that is controlling the fuel servo valve, thus the fuel rate to the combustion chamber is optimized [GCa76][WRo72]. The core of the system is represented by an electronic device whose general diagram is presented in Figure 1.4.

As shown in Figure 1.4, this electronic device is connected to two exhaust gas temperature sensors, to a pressure transducer (which collects data regarding the air pressure resulting from the compressor, necessary for combustion to take place), and finally to two tachometers that record the angular speed of the power turbine shaft [GCa76][RBo61][WRo72]. The output is connected to a servo valve that regulates the fuel flow to the combustion chamber.

The two tachometers are connected to electromagnetic transducers. The outputs of these tachometers are coupled to a maximum selector which will deliver the signal that has a greater value. The maximum selector is connected simultaneously to a main control circuit and a function generator. The working principle of this function generator is demonstrated in Figure 1.5. The function generator acts as an acceleration limitator during the starting

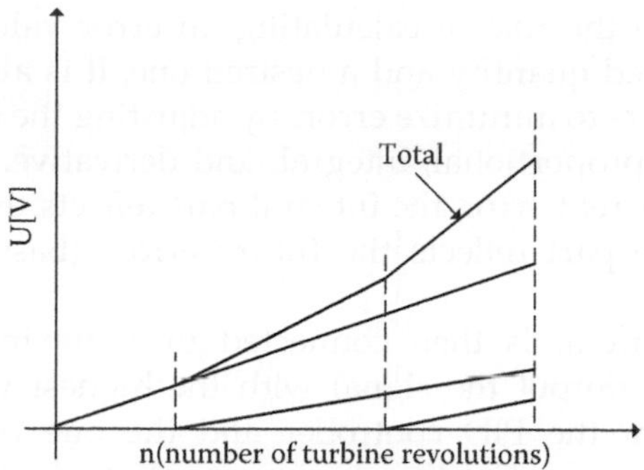


**FIGURE 1.5**
Approximate description of the operating principle for the FIAT function generator. (From G. Caire et al., Device for controlling gas turbine engines. U.S. Patent 3,987,620, issued October 26, 1976, available online: http://www.uspto.gov [accessed February 27, 2013]; R. Bodemuller et al., Gas turbine acceleration control. U.S. Patent 2,971,338 issued February 14, 1961, available online: http://www.uspto.gov [accessed February 28, 2013]; and W. Rowen et al., Constant power control system for gas turbine. U.S. Patent 3,639,076, issued July 14, 1972, available online: http://www.uspto.gov [accessed February 28, 2013]. With permission.)

sequence of the engine [GCa76][RBo61]. In other words, when for a certain reason the main control circuit output tends to exceed the limit imposed by the speed variator, the function generator intervenes [GCa76][WRo72].

The task of the main control circuit is to compare the voltage proportional to the engine speed resulting from the maximum signal selector with a voltage imposed from outside through a potentiometric speed variator [GCa76][RBo61]. The output of the main control circuit is an error signal. The role of the potentiometer connected to the feedback loop of this circuit is to obtain a steady operation that can be directly influenced from outside by an independent operator [ALo70][GCa76][WRo72].

Another important component of the electronic device controlling the fuel flow is the level corrector. Its aim is to provide a signal representing the admissible limit value of the output temperature corrected with respect to the pressure of the air resulting from the compressor and entering the combustion chamber [GCa76][RBo61][WRo72].

At the same time, the two exhaust gas temperature sensors supply an average value and are connected to a selector that is enabled only during the starting sequence of the motor by a control [ALo70][GCa76] which, for sake of simplicity, is not shown in Figure 1.4.

In the next stage, an OR logic gate simultaneously receives signals from the summing circuit and the selector and actuates a comparing circuit. The comparator generates an output error signal that represents the difference between the turbine temperature and the reference temperature resulting from the level corrector [GCa76][RBo61][WRo72]. This error signal enters the adjusting circuit which is in fact a PID (proportional-integral-derivative) controller. As is well known, such kinds of controllers are often used in industry and have the role of calculating an error value as a difference between a measured quantity and a desired one. It is also the function of these PID controllers to minimize errors by adjusting the inputs using three types of controls: proportional, integral, and derivative. The proportional part reflects the current error, the integral part reflects the previous errors, and the derivative part reflects the future errors (based on the current change rate).

The adjusting circuit is then connected to a maximum selector that will supply at the output the signal with the highest value between the one resulting from the PID controller and the one resulting from the potentiometer. The output of the maximum selector is further coupled to two lowest wins logic gates that control the fuel rate sent to the turbine (as a function of temperature) [ALo70][GCa76]. As described, the purpose of the lowest wins logic gate is to provide at the output the signal having the lowest value among the signals at the input.

The role of the maximal output limitator, on the other hand, is to provide a comparison between the output of the adjusting circuit and the voltage generated by an internal potentiometer of the limitator [GCa76][WRo72].

For further information on the fuel control system for the gas turbine engine developed by FIAT, refer to [ALo70][GCa76][RBo61][WRo72].

The car propelled by this gas turbine engine was called FIAT Turbina, had a power of 300 hp (approximately 224 kW), and achieved a maximum speed of 250 km/h. One of its most important characteristics was a very low drag coefficient of approximately 0.14, measured in the wind tunnel of Politecnico di Torino, Italy [MAT-- ]. The drag coefficient represents a dimensionless quantity and is employed to determine the resistance of an object in a fluid environment (like water or air). The first tests of this car have been carried out on the runway of the Caselle airport in Turin, Italy [MAT-- ].

## 1.3 The Fuel Control System for a Gas Turbine Engine Developed by Ford

In 1955, Ford became the first company to make public the results of its research about gas turbine engines for the automotive industry. The first step was very cautious. A regenerative 150 hp (approximately 112 kW) gas turbine was installed on a 1954 truck, but the results achieved turned out to be not very impressive, according to company reports [FMC-- ].

Seven years later, Ford developed a 300 hp (224 kW) gas turbine which led subsequently to the development of a 600 hp (448 kW) variant, commissioned by the Department of Defense. This new, improved model was used to power a super-transport prototype truck. This turbine engine offered important advantages such as easy startup in cold environments, reduced oil consumption, reduced emissions, low noise, instant fuel-power capability, few vibrations, as well as high torque at low speeds [FMC-- ].

This truck, driven across the United States, was rated at 600 hp (approximately 441 kW) and had a peak efficiency of 37%, which was much higher than those of the competing gas turbine engines available at that time [Robin Mackay, Agile Turbine Technology].

The fuel control system of this gas turbine engine was composed of a compressor and a turbine which could be either of radial or of axial flow [AFM69][AM69][APM74]. These two components could be simultaneously of the same type or of different types without influencing the system operation [AM69][APM74][THM64].

The main difference between the axial and radial turbines resides in the direction of the working fluid, compared to the turbine shaft. In the case of the radial turbine, the flow is diverted by the compressor at 90° toward the combustion chamber. As a result, the radial turbine is more efficient, more robust, and presents less thermal and mechanical wear. The disadvantage is represented by the fact that for applications that require above 5 MW of power, this type of equipment is no longer appropriate due to the costs and weight implied by the rotor. As mentioned above, the power of this motor was of 441 kW, and so this type of unit became suitable. On the other hand, in axial turbines, the working fluid flows practically parallel to the shaft, and this type of equipment is generally used for jet engines, high-speed ship engines, or for distributed generation.

For our particular case, the Ford compressor consists of a group of 12 inlet guide vanes (see Figure 1.6) which are equally spaced and mounted on a common actuator. The role of these inlet vanes is to control the air flow through the compressor. In this way, the needed horsepower is influenced by the torque control rather than air speed through the inlets [AFM69][AM69][APM74]. The turbine itself is driven by the products resulting from the combustion chamber.

Additionally, the fuel pump was designed in such a way that it supplied more fuel than required and thus the system never ran out of gas. As a consequence, a supplementary fuel pipe was connected to the supply line so that the fuel was bypassed back to the entry of the fuel pump through a measuring device [AM69][APM74].

Other important components of the engine were the heat exchangers which had to supply to the combustion chamber low-temperature air at high pressure resulting from the compressor discharge [AFM69][APM74] and also the reduction gear box. The latest is connected to the output shaft of the power turbine (see Figure 1.1). At the same time, this power reduction gear box serves also as an input [AM69][APM74] to a hydraulic torque converter. This converter consists of a turbine (suitable for rotation with the input element from the transmission system of the truck), a pump (suitable for rotation together with the shaft), and finally a variable angle stator.

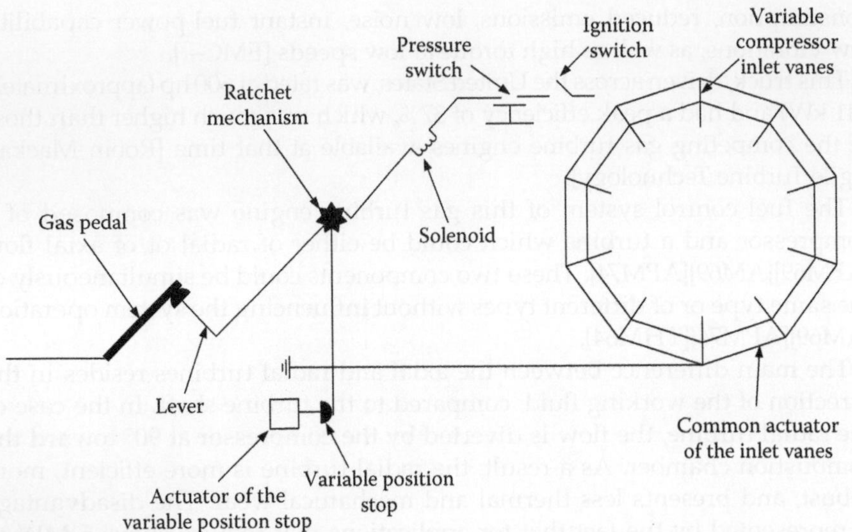


**FIGURE 1.6**
Approximate view of the fuel control system for the Ford gas turbine engine. (From A.F. McClean et al., Gas turbine control system. U.S. Patent 3,485,042 issued December 23, 1969, available online: http://www.uspto.gov [accessed February 25, 2013], and A.P. McClean et al., Gas turbine control system. U.S. Patent 3,795,104 issued March 5, 1974, available online: http://www.uspto.gov [accessed February 25, 2013]. With permission.)

The control system of the gas turbine implemented by Ford has three circuits:

- Movement of the compressor inlet guide vanes (see Figure 1.6)
- Movement of the stator vanes of the hydraulic torque converter (not shown in Figure 1.6)
- Monitoring with a fuel flow metering device the fuel entering the combustion chamber (not shown in Figure 1.6) [AM69][APM74]

The compressor inlet guide vanes provide an optimal air flow during the starting, acceleration, or idle operation of the engine, without exceeding the temperature limits. The stator vanes of the hydraulic torque converter assure the desired speed of the vehicle as well as the maximum operating temperature. Finally, the metering device imposes the optimal fuel quantity which enters the combustion chamber in such a way that both mechanical and temperature limitations are fulfilled.

Another important feature of this engine is that, in the case of the partial load operation, it is preferable to maintain the unit at approximately 55% of the full load so that accelerations from idle can be attained rapidly.

A similar situation exists in the case of the natural gas MTs used for DG. The major difference here is that it is important to keep them running at least at 50% of the full load, otherwise the pollutant emissions tend to become very high, according to experimental results (see Chapter 3).

As described in Figure 1.6, the core of the control system is represented by the adjustable compressor guide inlet vanes. These facilitate exactly this kind of operation, maintaining the unit in case of idling, at 55% of the full load [APM74]. Closing these vanes will result in a decrease of engine torque. During the starting sequence, the nozzle formed by the inlet vanes has to be open due to the fact that the air flow needs to be maximal. As can be observed in Figure 1.6, the adjustable inlet vanes have a common actuator that is connected through a linkage and a ratchet mechanism to the gas pedal of the truck. In this way, the closing and opening of the inlet vanes are directly influenced by the depression and the release of the pedal [AFM69][APM74][THM64].

The electrical circuit including the solenoid, the pressure switch, and the ignition switch contributes also to the opening of the compressor inlet vanes. Normally, the ignition switch is closed. Thus, the solenoid actuates a dog (for the sake of simplicity, not shown in Figure 1.6) to lock a part of the ratchet mechanism from a clockwise rotation [APM74][THM64]. Furthermore, during the starting sequence of the engine, when the above-mentioned nozzle needs to be opened wide for maximal air flow, the closing of the ignition switch together with the depression of the gas pedal will rotate the gas inlet vanes to the open position.

The temperature control described previously is performed through the variable position stop and the corresponding actuator. Sometimes there are situations when the operating temperatures can be exceeded. The solution to this problem would be to open the guide inlet vanes [AFM69][APM74]. In this case, the variable position stop driven by an electric motor (actuator) will move the vanes into the open position. The electric motor is connected to an amplifier (again for the sake of simplicity, not shown in Figure 1.6) whose inputs are represented by two signals: the actual speed and the actual temperature. When the temperature becomes too high, the actuator will drive the variable position stop to the right and so the compressor guide inlet vanes will be opened [APM74][THM64]. After the temperature decreases to the desired level, the actuator will come back to its initial position.

For further reading and more detailed information on the control system for a gas turbine engine implemented by Ford, please refer to [AFM69][AM69][APM74][THM64].

The truck powered by this gas turbine engine was running on No. 2 diesel fuel which shows good performance in cold environments and was capable of pulling two trailers of 12.2 m in length at a speed of approximately 113 km/h.

## 1.4 The Fuel Control System for a Gas Turbine Engine Developed by Chrysler

The system implemented by Chrysler uses the same general working principle as the system implemented by Ford. For efficiency purposes, the air resulting from the compressor is preheated by the regenerator [AMP54][Cha65][Wat47] and driven to the combustion chamber where the fuel is added and finally burned. The combustion products are then directed through the two-staged rotor to drive the same while the exhaust gases are diverted toward the regenerator (in order to heat it) and, in the end, exhausted to atmosphere.

In the case of two rotor stages, these rotate separately from each other, the first stage having the role of actuating the compressor and the regenerator while the second stage is used to move the vehicle [Cha65][RH54][Wat47].

The basic idea of the control system implemented by Chrysler is to keep an optimum temperature in the combustion chamber. Another important aspect is to assure rapidly an increased fuel rate when increased power is required from the engine. In such conditions, the increased power is diverted initially to the first rotor stage which drives the compressor and, in this way, to accelerate it in order to provide the desired increased fuel quantity [AMP54][Cha65][RH54][Wat47]. That is why the most important part of the whole control system is represented by the fuel metering devices: an adjustable speed fuel throttle valve, a pressure responsive fuel valve, and a fuel scheduling assembly (see Figure 1.7) [Cha65][Wat47].

The throttle valve has the role of remaining fully open for a maximum fuel quantity during the acceleration of the air compressor and to decrease the fuel flow when the compressor has already achieved an initially imposed speed [Cha65]. The pressure-responsive fuel valve has to discharge the air pressure of the compressor in order to increase the fuel flow during the pressure increase of the combustion supporting air [Cha65][Wat47]. The core of the fuel metering system is represented by the fuel scheduling assembly and is fully investigated below.

The fuel scheduling assembly is depicted in Figure 1.7. This assembly receives fuel from the adjustable fuel throttle valve through the throttle assembly discharge conduct. As can be observed in Figure 1.7, this part has a high-pressure fuel inlet chamber (symbolized through the lower fuel buffer chamber) which communicates with the discharge conduct as well as with the fuel discharge chamber (upper fuel buffer chamber) through the metering orifice. This takes place while the engine is running. At the same time, the metering orifice is also connected to the metering rod and is filled by the latter (as described in Figure 1.7) when the engine is turned off [Cha65][RH54][Wat47].

The opposite ends of the metering rod are connected to two sealing bushings (not shown in Figure 1.7) which close the two fuel buffer chambers.

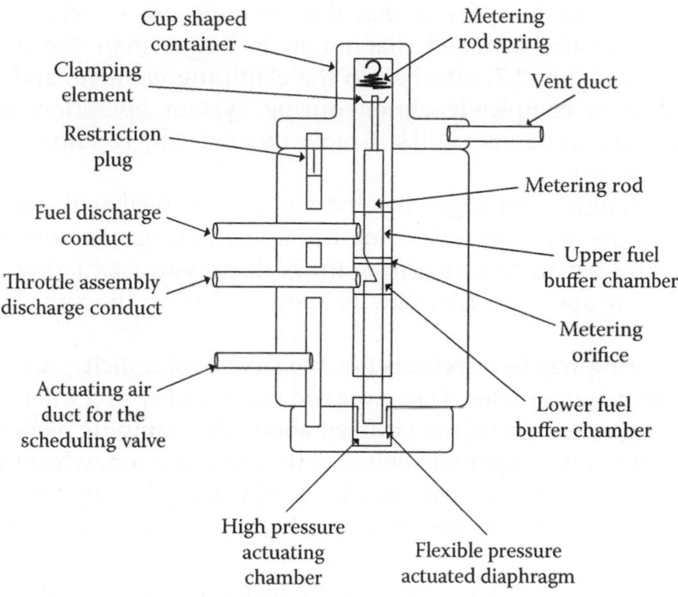


**FIGURE 1.7**
Approximate view of the fuel scheduling assembly for the Chrysler gas turbine engine. (From A.M. Prentiss, Fuel and speed control apparatus. U.S. Patent 2,691,268 issued October 12, 1954, available online: http://www.uspto.gov [accessed February 25, 2013]; A. Chadwick, Fuel control system for a gas turbine engine. U.S. Patent 3,183,667, issued May 18, 1965, available online: http://www.uspto.gov [accessed February 25, 2013]; and E.A. Watson et al., Liquid fuel pump governor. U.S. Patent 2,429,005 issued October 14, 1947, available online: http://www.uspto.gov [accessed February 25, 2013]. With permission.)

The role of the flexible pressure actuated diaphragm is to separate the high-pressure actuating chamber from the rest of the part [Cha65][RH54].

To reduce the pressure of the chamber in which the metering rod spring is located with respect to the pressure in the high actuating pressure chamber, a restriction plug will be used. When the compressor discharge increases, the pressure in the high-pressure actuating chamber will also increase, and so the metering rod is pushed upward against the metering rod spring [Cha65][Wat47]. As shown in Figure 1.7, the metering rod is equipped with an axially tapered metering interval which together with the metering orifice is capable of gradually increasing the fuel flow from the lower fuel buffer chamber to the upper fuel buffer chamber in case of a rod movement identical to the one described previously [AMP54][Cha65][Wat47]. When the compressor discharge pressure increases for supplying a greater quantity of combustion air, the fuel flow from the throttle assembly discharge conduct (through which the fuel enters the assembly from the throttle valve) to the fuel discharge conduct (through which the fuel exits the assembly) is also increased [AMP54][Cha65][Wat47].

Another important aspect is that the effective cross-sectional area of the flexible pressure actuated diaphragm is larger than the diaphragm (not shown in Figure 1.7) attached to the clamping element, and so when the vent duct is completely closed during system operation, the high-pressure actuating chamber will still move the metering rod upward against the spring [Cha65].

During the engine starting, when the fuel flow as well as the compressor air discharge are low, the rod spring has a low spring rate and is already compressed by the air resulting from the high-pressure actuating chamber. In this way, an optimum fuel flow is assured during the vehicle starting sequence.

The rod spring has two sections (for the sake of simplicity, not shown in Figure 1.7) of different rates. The lower end of the rod spring inferior section (which has a greater spring rate) is fixed above the clamping element and is activated to resist the upper movement of the metering rod, when the engine load surpasses the idling conditions [Cha65][Wat47]. Finally, the release of the optimum fuel flow toward the gas turbine takes place through the vent duct.

For the sake of simplicity, some of the elements of the fuel scheduling assembly are not depicted in Figure 1.7. For further information regarding this part of the fuel control system, please refer to [AMP54][Cha65][RH54][Wat47].

Using a fuel metering system with the configuration described above, a very effective response to the temperature and pressure of the inlet air is obtained which further enables an improved fuel flow control. The operation of the fuel control system can be seen in Figures 1.8 and 1.9.

If the lever of the fuel throttle valve is moved to idle (for acceleration purposes), the fuel flow will increase according to the upper solid curve in Figure 1.8, determined simultaneously by the metering rod and the throttle valve. When the compressor speed attains a value corresponding to the steady-state operation, the fuel flow will decrease very quickly along curve 1 to idle fuel level 5 on the dotted curve [Cha65][Wat47] in Figure 1.8. Curves 2, 3, and 4 together with the idle fuel levels 6, 7, and 8 correspond to different settings of the aforementioned lever [Cha65][Wat47].

The relation between the compressor pressure and fuel flow is illustrated in Figure 1.9. As can be seen in this figure, the fuel flow toward the combustion chamber increases with decreasing temperature of the inlet gases to the burner [AMP54][Cha65][Wat47]. For instance, during the starting sequence, when the engine is cold, the vent orifice of the fuel scheduling assembly will be fully open, thus the reaction air in the spring chamber (see Figure 1.7) will be at its lowest value and the pressure in the high-pressure chamber will move the metering rod upwards against the spring. When the operating conditions stabilize and the inlet gas temperature increases, the pressure in the spring chamber (see Figure 1.7) will increase, moving the metering rod downwards and hence reducing the fuel flow [Cha65]. In this way, not

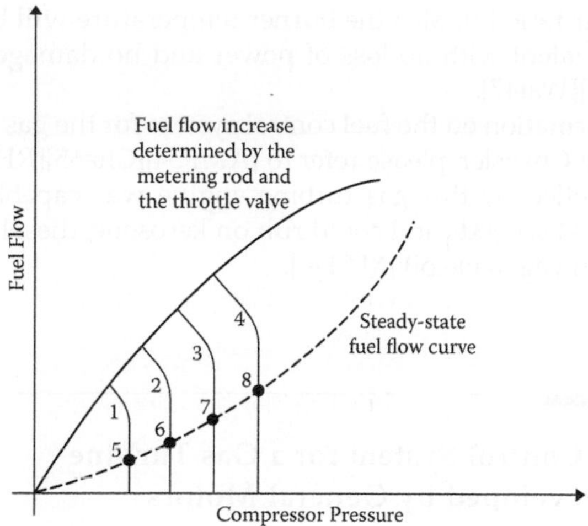


**FIGURE 1.8**
Approximate view of the operation characteristics for the Chrysler fuel control system. (From A.M. Prentiss, Fuel and speed control apparatus. U.S. Patent 2,691,268 issued October 12, 1954, available online: http://www.uspto.gov [accessed February 25, 2013]; A. Chadwick, Fuel control system for a gas turbine engine. U.S. Patent 3,183,667, issued May 18, 1965, available online: http://www.uspto.gov [accessed February 25, 2013]; and E.A. Watson et al., Liquid fuel pump governor. U.S. Patent 2,429,005 issued October 14, 1947, available online: http://www.uspto.gov [accessed February 25, 2013]. With permission.)

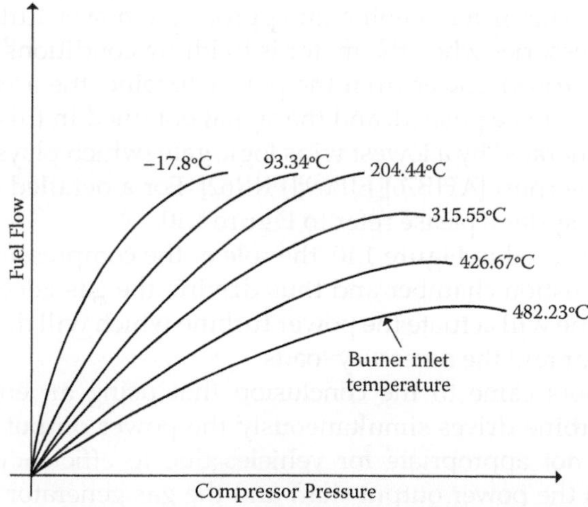


**FIGURE 1.9**
Approximate representation of the interdependency between the compressor pressure and the fuel flow. (From A. Chadwick, Fuel control system for a gas turbine engine. U.S. Patent 3,183,667, issued May 18, 1965, available online: http://www.uspto.gov [accessed February 25, 2013]; and E.A. Watson et al., Liquid fuel pump governor. U.S. Patent 2,429,005 issued October 14, 1947, available online: http://www.uspto.gov [accessed February 25, 2013]. With permission.)

only the fuel–air ratio but also the burner temperature will be at optimum, this being equivalent with no loss of power and no damage to the burner [AMP54][Cha65][Wat47].

For other information on the fuel control system for the gas turbine engine implemented by Chrysler, please refer to [AMP54][Cha65][RH54][Wat47].

The car propelled by this gas turbine engine was capable of achieving 97 km/h in just 14 seconds and could run on kerosene, diesel fuel, unleaded gasoline, or even vegetable oil [MTT-- ].

## 1.5 The Fuel Control System for a Gas Turbine Engine Developed by General Motors

The fuel control system implemented by General Motors presents an important advantage, namely the fact that the motor accessories like the alternator (for charging the car battery), the air conditioning, the oil and the fuel pumps or the power steering pump are driven by the power turbine (which actuates also the vehicle wheels) while the gas generator turbine rotates independently from the power turbine shaft and is free of any other accessory loads [AHB76][EJB69][FRR62].

Like the other fuel control systems, this one also responds to the power turbine speed that regulates the fuel flow to the gas generator [AHB76][EJB69][JAK73]. That is for keeping an appropriate power turbine speed for driving the accessories when the motor is in idling conditions.

To obtain increased power from the power turbine, the accelerator pedal of the vehicle has to be pushed, and the signal obtained in this way is added to the signal generated by a lowest wins logic gate (which plays the role of an underspeed governor) [AHB76][EJB69][FRR62]. For a detailed description of the fuel control system, please refer to Figure 1.10.

As can be observed in Figure 1.10, the role of the compressor is to provide air to the combustion chamber and thus driving the gas generator. The gas generator turbine will actuate the power turbine which will drive further the wheels of the car and the accessory loads.

General Motors came to the conclusion that using an engine where a single-shaft turbine drives simultaneously the power output shaft and the compressor is not appropriate for vehicles due to efficiency reasons. For instance, when the power output shaft and the gas generator turbine rotate independently, it is possible for the output shaft to come to a standstill position while the gas generator continues to function, and hence the engine torque is more efficient [AHB76][FRR62][JAK73].

Generally, the gas turbine engine compared to a normal piston engine is less efficient because the former has a slower response to the full power requirement [AHB76][JAK73]. Theoretically, one second is needed for the

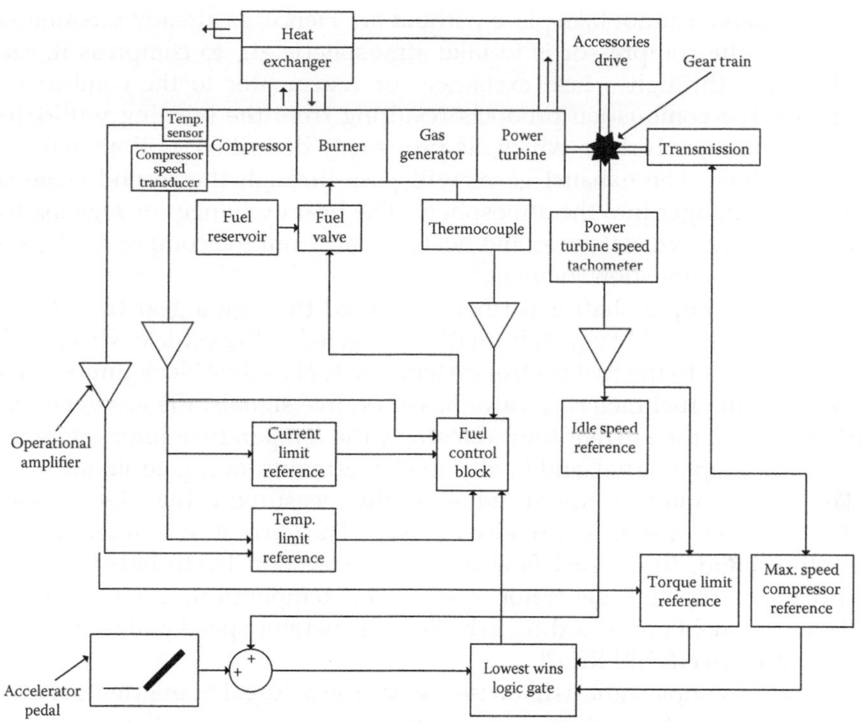


**FIGURE 1.10**
Approximate representation of the General Motors fuel control system for a gas turbine engine, (From A. Bell III et al., Automotive gas turbine control. U.S. Patent 3,999,373, issued December 28, 1976, available online: http://www.uspto.gov [accessed March 1, 2013]; E.J. Bevers et al., Turbine governor. U.S. Patent 3,439,496, issued April 22, 1969, available online: http://www.uspto.gov [accessed March 3, 2013]; and F.R. Rogers et al., Fuel control apparatus for a combustion engine. U.S. Patent 3,050,941, issued August 28, 1962, available online: http://www.uspto.gov [accessed March 3, 2013]. With permission.)

gas generator to achieve the full speed (from idle), and unfortunately the power turbine is not capable of functioning efficiently until the gas generator turbine reaches its full speed.

A major advantage of the engine developed by General Motors is the fact that the above-mentioned accessory loads are decoupled from the gas generator, and thus all of the power of the gas generator is diverted to the power turbine [AHB76][EJB69][FRR62]. In this way, the acceleration characteristic is very much improved. A direct consequence of this situation would be the prevention of the vehicle to creep.

One of the key components of the fuel control system is the fuel valve (see Figure 1.10) which is also capable of measuring the fuel flow. This valve is connected simultaneously to the fuel reservoir and to a fuel control block which imposes how much the valve will open and thus how much fuel will enter the combustion chamber where the burner is located. Combustion, on the other hand, cannot take place without air. Hence, as already mentioned, the role of the compressor is to take atmospheric air, to compress it, and to deliver it through a heat exchanger or regenerator to the combustion chamber. The combustion products resulting from the burning will drive the gas generator turbine which, as previously described, will actuate the power turbine. The exhaust gases will pass through the second stage of the heat exchanger into the atmosphere. The heat exchanger or regenerator cools down the exhaust gases and heats the air from the compressor before entering the combustion chamber.

The power output shaft is further connected through a gear train to the transmission assembly which is finally connected to the vehicle wheels.

Coming back to the fuel control system, the fuel control block imposes the opening of the fuel metering valve based on five signals: the gas generator turbine speed, the current limit reference, the temperature limit reference, the thermocouple signal, and finally, the lowest wins logic gate signal.

The gas generator turbine speed is usually measured using a tachometer, but other types of sensors can also be used. The current limit reference, on the other hand, limits fuel flow during acceleration [AHB76][JAK73] and responds to the gas generator speed. The temperature limit reference reduces the fuel flow according to both gas generator speed and motor inlet temperature [AHB76][FRR62].

The thermocouple signal is likewise of electrical nature. The inlet temperature of the gas generator turbine is measured, transformed in DC voltage, and through an operational amplifier fed to the fuel control block. In addition to the amplification of the aforementioned DC voltage, the operational amplifier must have a compensation circuit for the temperature lag [AHB76].

The lowest wins logic gate signal is generated choosing the lowest value between the torque limit reference, the gas generator speed, and the required acceleration of the car driver. The signal resulting from the torque limit reference is generated based on the inlet temperature and the speed of the power turbine. The maximum limit reference of the gas generator speed is set up through a potentiometer (for the sake of simplicity, not shown in Figure 1.10) [AHB76].

The idle speed reference is generated based on the tachometer signal from the power turbine speed and on two potentiometers (for the sake of simplicity, not shown in Figure 1.10) imposing the minimal values for the gas generator and the power turbine speeds [AHB76][JAK73]. Due to the sometimes long wiring paths between the system blocks, the use of operational amplifiers is appropriate to improve signal quality and thus obtaining an optimum control of the fuel flow from the reservoir to the combustion chamber. For a better understanding of the whole fuel control system, its operation can be synthesized using the plot in Figure 1.11. In the figure, $T$ represents the engine inlet air temperature.

For other detailed information on the General Motors fuel control system for a gas turbine engine, please refer to [AHB76][FRR62].

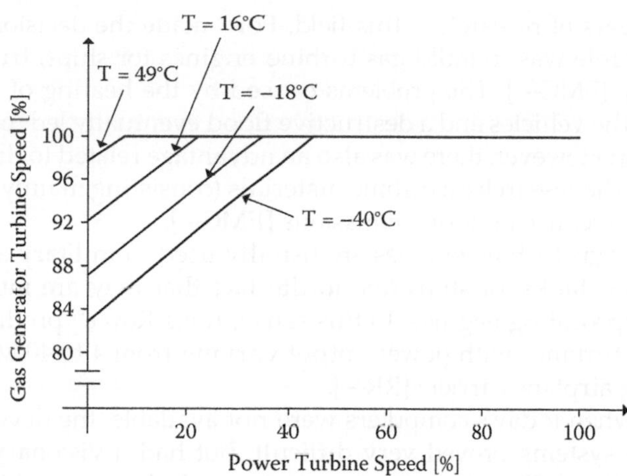


**FIGURE 1.11**
Approximate representation of the operation characteristics for the General Motors fuel control system. (From A. Bell III et al., Automotive gas turbine control. U.S. Patent 3,999,373, issued December 28, 1976, available online: http://www.uspto.gov [accessed March 1, 2013]; E.J. Bevers et al., Turbine governor. U.S. Patent 3,439,496, issued April 22, 1969, available online: http://www.uspto.gov [accessed March 3, 2013]; and J.A. Karol, Actuating device for a gas turbine engine fuel control. U.S. Patent 3,733,815, issued May 22, 1973, available online: http://www.uspto.gov [accessed March 3, 2013]. With permission.)

The concept cars equipped with this kind of gas turbine engine were capable of speeds of more than 160 km/h and ran on fuel oil, gasoline, or kerosene [CKS-- ][GMH-- ]. For further reference on the GM fuel control system for a gas turbine engine, please refer to [AHB76][EJB69][FRR62][JAK73].

Another manufacturer of gas turbine vehicles was Austin in the 1950s [AUS-- ]. Like the other manufacturers of gas turbine engines, Austin was confronted with the same issues, one of these being engine size. Due to this, it was decided to install this motor on a Sheerline model. The research was soon discontinued because of the other problems related to noise and high fuel consumption. For additional information on the Austin gas turbine engine, please refer to [AUS-- ].

In the 1950s Boeing® developed a gas turbine engine for trucks that was short lived as well. With time, the use of gas turbine engines turned out to be not very efficient for the automotive industry. This was mainly due to the costs involved.

Initially, it was believed that optimizing and improving components like regenerators, compressors, or burners would solve the problem related to the production of such equipment. But in the end, the costly materials necessary to resist the high temperatures generated by these engines and the high fuel consumption during idle operation forced these vehicles to be only prototypes [FMC-- ].

After 18 years of research in this field, Ford made the decision to open a plant whose role was to build gas turbine engines for ships, trucks, buses, and industry [FMC-- ]. The problems caused by the heating of the turbine installed on the vehicles and a destructive flood eventually led to the closing of this facility. However, there was also an advantage related to the operation of this plant–the research on turbine materials (consisting mainly of ceramic) that helped to reduce polluting emissions [FMC-- ].

At present, gas turbine engines are usually used in military applications like jet planes, tanks, or ships due to the fact that they are much smaller than the reciprocating engines. In this sense, Rolls Royce® produces a wide range of gas turbines with power output varying from 4 to 40 MW, capable of propelling airplane carriers [RR-- ].

In an era when today's computers were not available, the development of such control systems proved very difficult, but had a visionary character. In the end, the difficulties were overcome and the gas turbine engines designed for vehicle propulsion came to represent an important chapter in the evolution of natural gas MTs. In this way, the foundation has been laid of one of the most efficient sources in the DG.

# 2

# *Natural Gas Microturbines in Distributed Generation*

After a hiatus of approximately 30 years, gas microturbines (MTs) again caught the attention of the scientific community—this time as a source for distributed generation (DG). The first ideas about this appeared in the late 1980s, but the first units were not commissioned until the late 1990s.

During this time, the players in this market changed. For instance, the former Swedish company Turbec AB® is now producing MTs in Italy, while Elliott Energy Systems, Inc.® was acquired by Calnetix Inc.® in 2007 and in 2010 was finally acquired by Capstone because the design of the Calnetix MT was compatible with the product line from Capstone, and it filled the gap between the C65 and C200 developed by Capstone [CPT-- ]. The Calnetix MT has a rated power of 100 kW, while C65 and C200 present rated powers of 65 kW and 200 kW, respectively. Capstone [Ha03][So07] offers very good equipment due to the fact that the bearings of the MT require no lubrication and thus the overall efficiency, and the electrical efficiency in particular, is very much improved. These two types of natural gas MTs are thoroughly investigated here.

As stated in Chapter 1, a gas MT (and also the Capstone turbogenerator) comprises the following key components: the engine which can be driven either on liquid fuel or gas, the fuel system including the gas boost compressors (which have the role of feeding the MT with gas at an appropriate pressure), and the power converter providing electrical energy at 50 Hz or 60 Hz, depending on the application (which will be thoroughly described in Chapter 4).

The electrical performances of models C30 and C65 can be observed in Figures 2.1 and 2.2. Cross sections of these two MTs are shown in Figures 2.3 and 2.4.

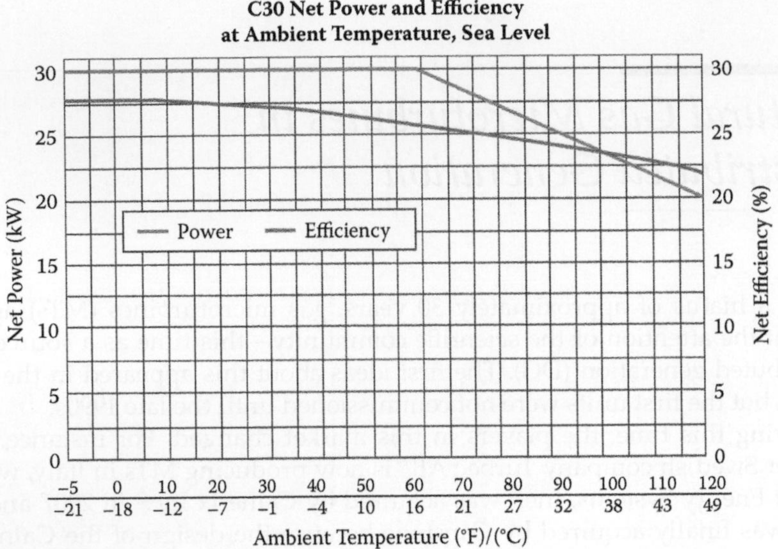


**FIGURE 2.1**
The electrical net power and net efficiency of a C30 MT as a function of temperature. (Courtesy: Capstone Turbine Corporation.)

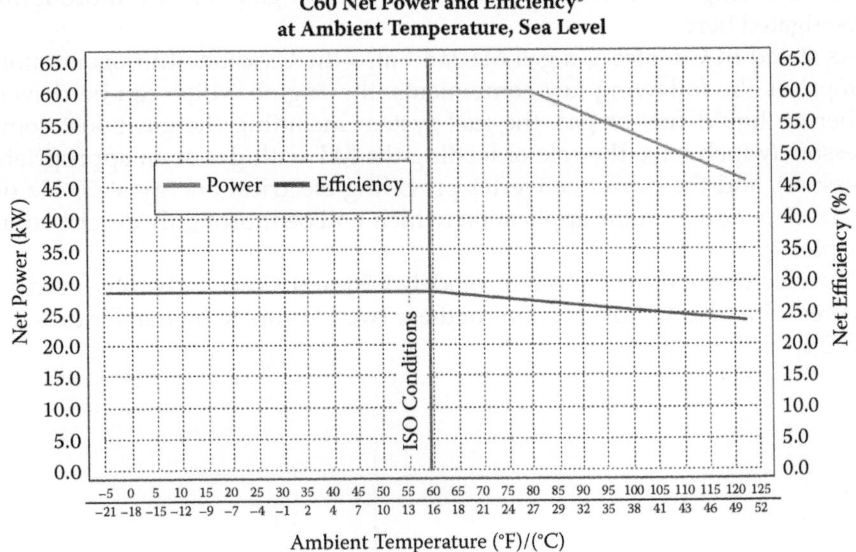


**FIGURE 2.2**
The electrical net power and net efficiency of a C65 MT as a function of temperature. (Courtesy: Capstone Turbine Corporation.)

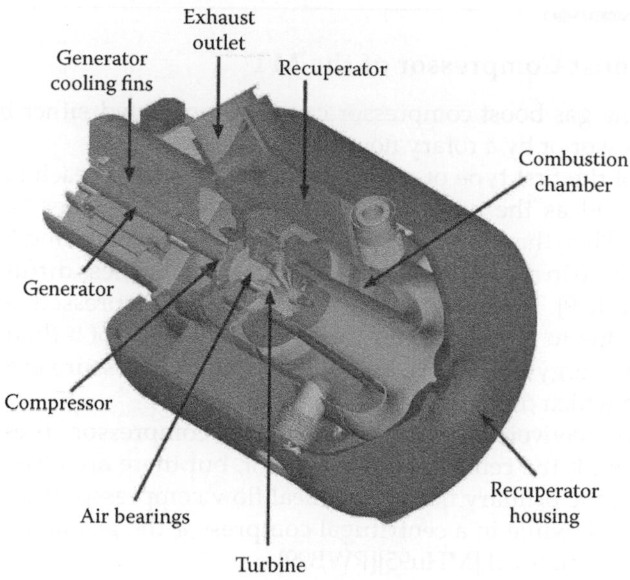


**FIGURE 2.3 (See color insert.)**
Cross section of a C30 natural gas MT. (Courtesy: Capstone Turbine Corporation.)

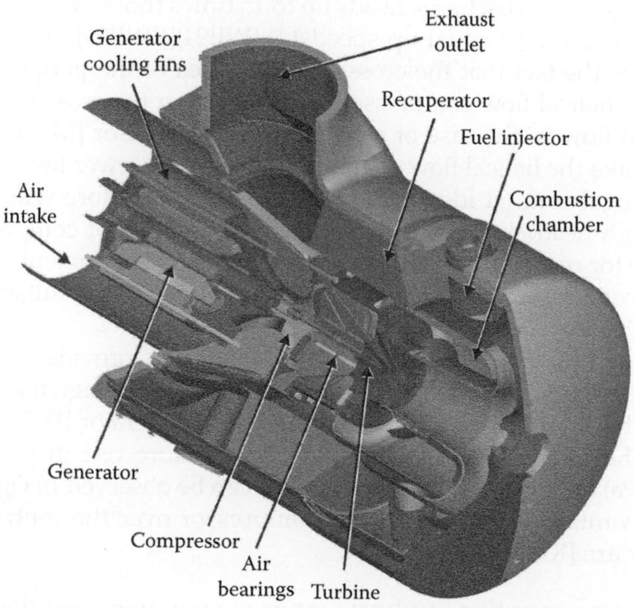


**FIGURE 2.4 (See color insert.)**
Cross section of a C65 natural gas MT. (Courtesy: Capstone Turbine Corporation.)

## 2.1 Gas Boost Compressor of the MT

The Capstone gas boost compressor can be represented either by a helical flow compressor or by a rotary flow compressor.

The role of the first type of compressor is to provide to each fluid particle a velocity head as the particle passes through the impeller blades of the compressor. Then the conversion between the velocity head and the pressure head takes place in a stator channel which acts as a vaneless diffuser [BGD95][HWB94][RWB99]. The velocity head of a fluid can be expressed as the energy of the fluid due to its bulk motion. The pressure head of a fluid represents the internal energy of that fluid resulting from the pressure exerted on the container in which the fluid is kept.

Given this conversion, the helical flow compressor presents some similarities with the centrifugal compressor, but there are also differences. For instance, the primary flow in a helical flow compressor is asymmetrical and peripheral, while in a centrifugal compressor the primary flow will be radial and symmetrical [MHu95][RWB99].

The flow pattern in a helical compressor permits the flow particles to pass through the impeller blades many times, each time acquiring kinetic energy [BGD95][HWB94][RWB99]. At the next stage, the fluid particles will reenter the adjacent stator channel where their kinetic energy will be converted into potential energy. The large number of passes makes the helical flow compressor generate discharge heads up to 15 times those produced by a centrifugal compressor (at equal tip speeds) [HWB94][RWB99]. Another important aspect is the fact that the cross-sectional area of the peripheral flow in the case of a helical flow compressor is smaller than the cross-sectional area of the radial flow in the case of a centrifugal compressor [MHu95][RWB99]. This will make the helical flow compressor operate at lower flows than those of a centrifugal unit (at identical tip speeds) and therefore will result in a low-flow high-head characteristic that makes this type of compressor more appropriate for certain applications where a rotary displacement compressor, low-speed compressor, or reciprocating compressor are not suitable [BGD95][HWB94][RWB99].

The helical flow compressor can also be used to provide high-pressure working fluid to a turbine, dropping the fluid pressure through the turbine and extracting the shaft mechanical power through a generator [BGD95][MHu95][RWB99]. The interdependency between the pressure rise in psig (1 psig = 6894.75728 Pa) and the fluid flow rate in $m^3/s$ can be observed in Figure 2.5.

Other advantages of a helical flow compressor over the centrifugal flow compressor are [RWB99]:

- Surge-free operation in a broad range of operating conditions
- No product wear and oil contamination due to the fact that there are no lubricated surfaces or rubbings

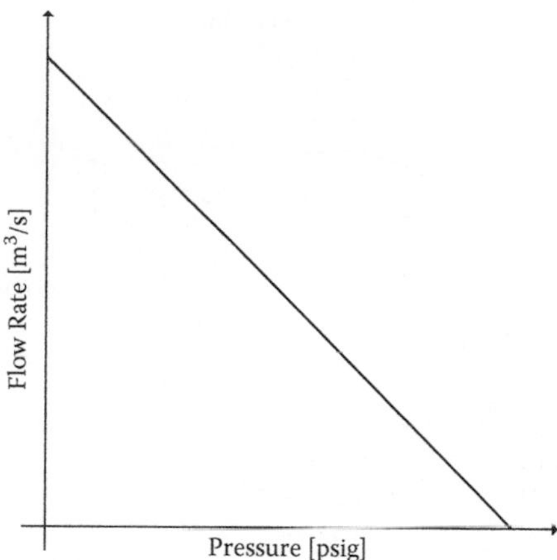


**FIGURE 2.5**
The approximate interdependency between the pressure rise in a helical flow compressor and the fluid flow rate. (From H.W. Brockner et al., Automotive fuel pump housing with rotary pumping element. U.S. Patent 5,310,308, issued May 10, 1994, available online: http://www.uspto.gov [accessed March 7, 2013]; M. Huebel et al., Peripheral pump, particularly for feeding fuel to an internal combustion engine from a fuel tank of a motor vehicle. U.S. Patent 5,468,119, issued November 21, 1995, available online: http://www.uspto.gov [accessed March 8, 2013]; and R.W. Bosley, Helical flow compressor/turbine permanent magnet motor/generator. U.S. Patent 5,899,673, issued May 4, 1999, available online: http://www.uspto.gov [accessed March 7, 2013]. With permission.)

- Lesser stages, compared to a centrifugal compressor
- Long life (approximately 40,000 hours) which is limited only by bearings
- Higher operating efficiencies

The back view of this helical flow compressor can be observed in Figure 2.6. The role of the fluid inlet is to provide working fluid to the compressor, while the role of the fluid outlet is to remove this fluid from the compressor.

Another important aspect is related to the ignition (which will be fully investigated in the next section). Normally, in order to start the MT, the helical flow compressor will have to run backward [RWB99]. This is to reduce the upstream pressure of the gas (usually supplied from a pipeline) entering the turbine. The gas fuel header pressure must be very low in order for ignition to take place [MHu95][RWB99].

When MT speed increases, the turbine discharge pressure will increase as well, and the gas pressure in the header that feeds the unit combustor must be maintained at a value greater than the pressure discharge value. So, for instance, if the gas pressure in the pipeline is 20 psig, for turning on the MT one has to reduce this pressure by 19 psig. After ignition has taken place,

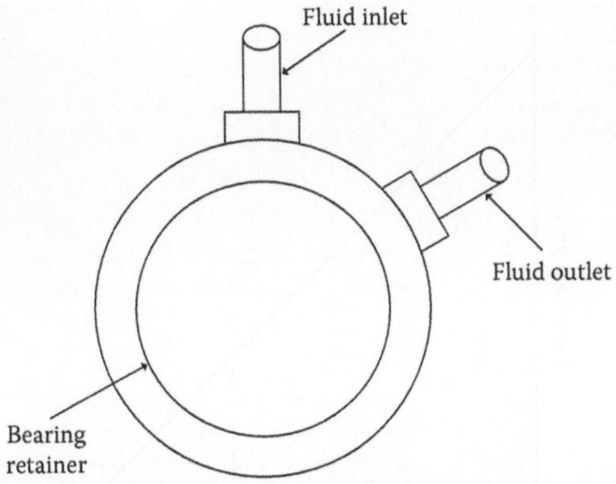


**FIGURE 2.6**
Schematic back view of the Capstone helical flow compressor. (From H.W. Brockner et al., Automotive fuel pump housing with rotary pumping element. U.S. Patent 5,310,308, issued May 10, 1994, available online: http://www.uspto.gov [accessed March 7, 2013]; M. Huebel et al., Peripheral pump, particularly for feeding fuel to an internal combustion engine from a fuel tank of a motor vehicle. U.S. Patent 5,468,119, issued November 21, 1995, available online: http://www.uspto.gov [accessed March 8, 2013]; and R.W. Bosley, Helical flow compressor/turbine permanent magnet motor/generator. U.S. Patent 5,899,673, issued May 4, 1999, available online: http://www.uspto.gov [accessed March 7, 2013]. With permission.)

header pressure can increase. Normally, ignition occurs when the helical flow compressor still runs backwards. More details about ignition can be found in the Section 2.2.

For more details on the helical flow compressor, please refer to [BGD95][HWB94][MHu95][RWB99].

The rotary flow compressor is used especially in urban areas where natural gas pressure is very low, because this type of compressor is capable of boosting the gas pressure from 0.2 psig to 55 psig [CPTC-- ]. Other advantages of the rotary flow compressor are:

- It is lubrication free
- It has improved maintenance costs
- It demonstrates increased efficiency
- It offers no liquid contamination risk of the fuel flow, which could result from the lubrication

There are two types of technology used for this compressor. One consists of ball bearings and the other consists of the well-known Capstone air bearings (which are also used for the shaft). Ball bearings are suited for applications in which the initial gas pressure is valued in the range of 5–15 psig, while air bearings are appropriate for gas pressures as low as 0.2 psig [CPTC-- ].

## 2.2 The Ignition System

The basic principle for turning on a gas MT is to continuously compress the inlet air which will be further mixed with fuel in an inflammable proportion, thus the whole mixture becomes capable of igniting itself from an appropriate source. The energy resulting from this process is then converted to rotary energy, which through the turbine will finally drive the electrical generator. The last stage of this process consists of releasing into the atmosphere the exhaust gases after these have already given some of their residual heat to the incoming inlet air.

Generally, a MT has to be accelerated by an external power source in order to supply sufficient air flow to the combustor for the lighting-off [ECE00][Hol74][JTM74]. In this way, engine speed will vary as a function of the starter motor speed as well as of the torque [ECE00][ESH74] [WCo74]. In such a case, fuel flow to the MT is determined using an "open-loop" approach depending on a variety of factors like atmospheric pressure or temperature [ECE00][WCo74]. There are situations in which turning on MTs at high altitudes and low temperatures proves to be problematic.

In the system proposed by Capstone, ignition takes place practically under any circumstances and does not depend on the above-mentioned factors. Why is that?

In the classical open-loop approach, ignition takes place only when the values of temperature and atmospheric pressure are precisely determined and achieved, in which case the proportions of the air–fuel mixture are considered to be ideal [ECE00][Hol74][JTM74]. The Capstone system works as shown in Figure 2.7. As can be observed, at a time I the MT will reach a constant speed (of approximately 14,000 rpm), and this will be maintained by the permanent magnet generator [ECE00]. This constant speed will be kept until after ignition [ECE00][ESH74][WCo74]. At that time I, the fuel flow is initiated and constantly ramped until it crosses the curve associated with the air flow. The point *P* where these two curves intersect represents the optimum fuel-to-air ratio, and this will occur at time II [ECE00][Hol74][JTM74]. The answer to the question above lies in this diagram. In other words, the intersection point *P* (regardless of the curve height associated to the air flow or of the slope of the fuel flow curve) is always achieved no matter the values of atmospheric pressure or temperature [ECE00][ESH74][Hol74].

At the moment III, when the exhaust gas temperature confirms the fact that ignition took place, the fuel flow will be scheduled according to the "closed-loop" principle and will depend on the MT acceleration and exhaust gas temperature [ECE00].

The unit that controls the fuel flow works according to the flowchart in Figure 2.8. The first block in the figure represents MT acceleration to 14,000 rpm, as stated previously. After this speed has been reached, the ignitor is turned on as it is also the fuel valve which is electrically driven (see also

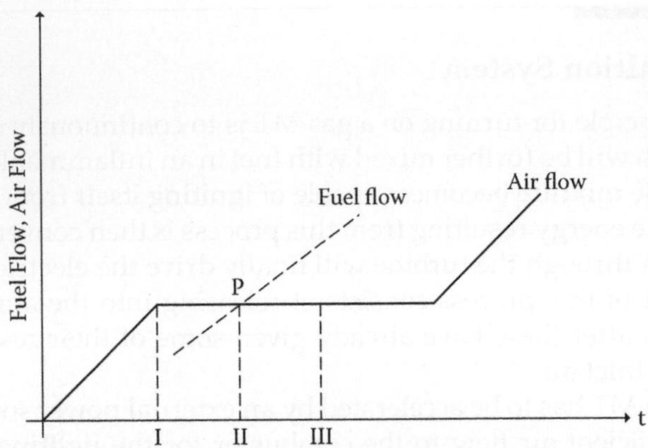


**FIGURE 2.7**
The approximate interdependency between fuel flow, air flow, and time. (From E.C. Edelman, Gas turbine engine fixed speed light-off method. U.S. Patent 6,062,016, issued May 16, 2000, available online: http://www.uspto.gov [accessed March 8, 2013]; and J.T. Moehring et al., Light-off transient control for an augmented gas turbine engine. U.S. Patent 3,834,160, issued September 10, 1974, available online: http://www.uspto.gov [accessed March 10, 2013]. With permission.)

Chapters 4 and 5) [ECE00][JTM74][WCo74]. The next step consists of fuel flow ramping (as in Figure 2.7). The decision block has the role of determining if the exhaust gas temperature of 10°C has been attained. If this has been attained, the MT is further accelerated (up to a speed of approximately 98,000 rpm) and the exhaust gas temperature control begins [ECE00]. If the temperature of 10°C has not been reached, the unit continues ramping the fuel flow until this temperature is achieved. When this temperature has been reached, ignition has taken place.

For more information on the Capstone ignition system, please refer to [ECE00][ESH74][Hol74][JTM74][WCo74].

## 2.3 The Shaft

After ignition has taken place, the shaft of the MT begins to rotate. Various parts of the MT, like turbines, fans, compressors, or generators, are connected to the shaft. When designing such a shaft, properties such as easy assembly and disassembly as well as minimal lubrication of the surfaces that come into contact are of great importance. The Capstone shaft is no exception to this rule. A graphical representation of this element is presented in Figure 2.9.

The Capstone shaft is a double diaphragm compound shaft [HEF81] [ILL73][PBV02]. It is made up of a rotatable shaft supported by two journal

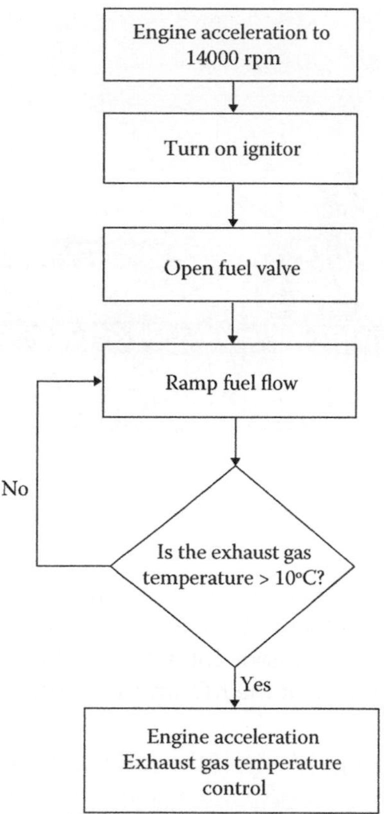


**FIGURE 2.8**
Flowchart of the fuel flow control unit. (From E.C. Edelman, Gas turbine engine fixed speed light-off method. U.S. Patent 6,062,016, issued May 16, 2000, available online: http://www.uspto.gov [accessed March 8, 2013]; and E.S. Harrisson et al., Gas turbine start-up fuel control system. U.S. Patent 3,844,112, issued October 29, 1974, available online: http://www.uspto.gov [accessed March 10, 2013]. With permission.)

bearings, a second shaft supported by a single journal bearing and also by a bidirectional thrust bearing, and a flexible disk shaft having two flexible disk diaphragms [GHh89][JNS92][PBV02].

An important aspect of this kind of design is that this shaft permits relatively large misalignments of the three journal bearings, given the flexible disk shaft. Other advantages are [JNS92][PBV02]:

- The capacity for the incorporation of boreless turbine rotors
- No lubrication required for the coupling surfaces between two rotor sections operating at high rotational speed
- Improved flexibility when performing maintenance operations

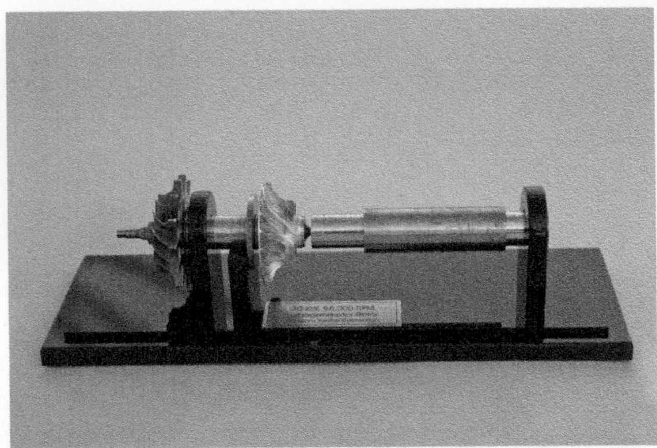

**FIGURE 2.9**
The shaft of the Capstone MT. (Courtesy: Capstone Turbine Corporation.)

- Minimization of loose components on the contact surface between two rotor sections which otherwise could become misaligned and sucked up in the compressor intake (such kind of ingestion could lead to serious damage of the MT and put it down for a long period of time)
- Ease of design change of any of the rotor sections

For more information on the Capstone shaft, please refer to [GHh89][HEF81][ILL73][JNS92][PBV02].

The most important component of the rotor would be represented by the air bearings, which assure practically no friction between the rotor itself and the other non-rotating parts of the MT. These bearings consist of a bushing, a shaft (rotating within the bushing), a layer of fluid (usually air) between the bushing and the rotor, and non-rotating compliant foil members which are connected to the bushing through a number of undersprings. For a graphical representation of this air bearing, please refer to Figure 2.10 [ASV68][ASV79][DHW06].

The foil elements form a certain number of wedge-shaped channels that converge in thickness toward the direction of the turbine rotation [DHW06][DJM68][RGk72]. The viscous drag forces resulting from the turbine rotation will cause the fluid (in our case the air) to enter these channels. If the shaft rotates toward the bushing, the convergence angle of the wedge channels will increase, and thus the fluid pressure in the channels will increase [ASV79][DHW06][RGk72]. If the shaft moves in the opposite direction, the fluid pressure along the wedge channels will decrease. In this way, the air in the wedge channels will generate varying restoring forces on the rotor, preventing any contact between the shaft and the non-rotating parts of the MT during the operation [ASV68][DHW06][RGk72].

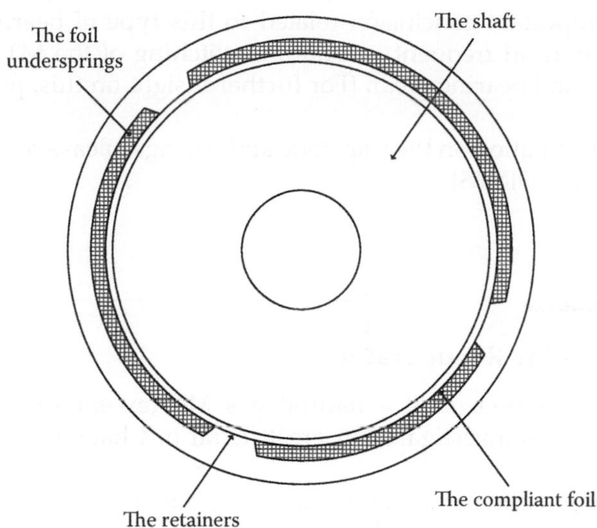


**FIGURE 2.10**
The Capstone air bearing—a schematic view. (From A. Silver et al., Selectively pressurized foil bearing arrangements. U.S. Patent 3,366,427, issued January 30, 1968, available online: http://www.uspto.gov [accessed March 12, 2013]; A. Silver et al., Foil bearing. U.S. Patent 4,178,046, issued December 11, 1979, available online: http://www.uspto.gov [accessed March 14, 2013]; and D.H. Weissert, Compliant foil fluid film radial bearing. U.S. Patent RE39,190, issued July 18, 2006, available online: http://www.uspto.gov [accessed March 12, 2013]. With permission.)

Continuous flexing of the metal foils will cause Coulomb damping of any axial movement of the rotor [DHW06][DJM68]. Coulomb damping represents a type of mechanical damping in which the energy is absorbed through sliding friction.

At low rotational speed or when the MT is switched off, the rotor comes into contact with the metal foils. This provokes a certain wear. Only at lift-off speeds is the air gap between the shaft and the non-rotating parts assured.

The role of the undersprings is to preload the fluid (air) foils against the shaft and to generate a certain dynamic stability [ASV79][DHW06][RGk72]. The bearing starting torque (which is ideally 0) is proportional to these preload forces [DHW06][RGk72]. They also increase the shaft speed at which the lift-off forces are capable of lifting the moving element. An important aspect related to this bearing is that its bore can be cylindrical or not (despite the fact that the rotor is always cylindrical).

The compliant foils seen in Figure 2.10 do not need to be three in number. The same is valid also for the foil undersprings [ASV68][ASV79][DHW06]. Five can also be suitable for this application. These undersprings do not present any special requirements. They can have a certain radius or be a rectangular sheet, or they can have various spring rates [ASV79][DHW06][DJM68].

The most important conclusion related to this type of bearing and unit operation is to avoid frequent on-and-off switching of the MT as this will result in rotor and bearing wear. (For further insight on this, please refer to Chapter 3.)

For more information on the Capstone air bearings, please refer to [ASV68][ASV79][DHW06][DJM68].

## 2.4 The Annular Recuperator

The next key component of a natural gas MT (except for the electrical generator, which is investigated in great detail in Chapters 4 and 5) is the recuperator.

Normally, the combustor, the compressor, and the turbine are located within this recuperator [YKg05]. For a cross section of the Capstone annular recuperator, refer to Figure 2.11. All of these components are usually located within the inner diameter of the recuperator, because in this way the heat of the exhaust gases can be reused and further transmitted to the air entering the combustion chamber. Thus, the overall efficiency of the MT will be very much improved, as will the fuel consumption. That is why it can be said that the main role of the regenerator is to optimally transfer the heat of the exhaust gases to the combustion air.

The present annular regenerator consists of a number of cold and hot cells. Cold cells are symbolized in Figure 2.11 through radial solid lines, while hot cells are shown through the spaces between these radiuses. In reality, the number of hot and cold cells is much larger than shown in Figure 2.11. For

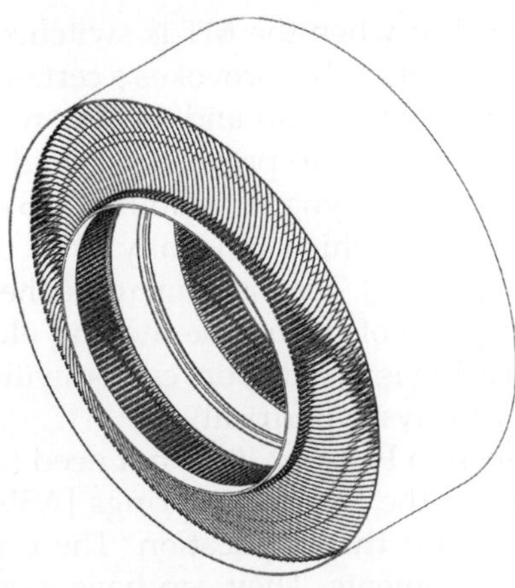

**FIGURE 2.11**
The Capstone annular recuperator. (Courtesy: Capstone Turbine Corporation.)

the sake of simplicity and for better understanding, this number has been greatly reduced.

Each of the cold cells has a fluid inlet in the inner diameter, located in the vicinity of the second end, and a fluid outlet in the annular housing, located near the first end [DGB91][WRR02][YKg05]. The hot cells alternate with the cold cells and allow the hot fluid to pass from the first end to the second end of the outer diameter [YKg05].

Cold cells are formed in such a way that the fluid inlet and outlet are diagonally opposed in order to even the fluid flow, and above all, the fluid flow resistance [JNa74][SHm53][YKg05]. The cold cells present a curved configuration because they have a rectangular cross section.

To further improve the heat transfer, there are also some protuberances that make the connection between the cold and hot cells. Due to the fact that these two types of cells come into contact with one another, the heat from the exhaust gases is transferred to the air (initially cold) entering the combustion chamber.

For more information on the Capstone annular recuperator, please refer to [DGB91][JNa74][SHm53][WRR02][YKg05].

## 2.5 Catalytic Reactor for Pollutant Emissions Minimization

Natural gas burning results not only in the production of thermal and mechanical energy but also in the emission of unwanted pollutants like $NO_x$ (nitrogen oxides), CO (carbon monoxide), VOCs (volatile organic compounds), and THCs (total hydrocarbons).

The main purpose for maintaining a low level of pollutant emissions is to keep an appropriate AFR (air-to-fuel ratio). This ratio is directly influenced by an air valve located at the compressor side and also by a digital power controller that controls the opening of this valve. An ideal AFR has to be constant for the fuel and power flow range of the turbogenerator [BDy02][JDe74].

Another important aspect of the Capstone system is the software that optimizes the operation of this controller. This software regulates the AFR value in the primary zone of the combustor. This regulation takes place based on the values stored in this computer program which consist of at least one AFR value and a set of values for fuel flow and air control [BDy02]. Figure 2.12 gives an illustration of this.

Theoretically, the air valve as well as the fuel valve in Figure 2.12 can be controlled by a separate analog controller [BDy02]. Usually, this is not the case due to the cost that would arise in this situation.

The startup sequence of the MT begins with the air entering the compressor through the air filter. In most cases, the air pressure is identical to room pressure. Then this air is compressed here to a higher pressure and will exit

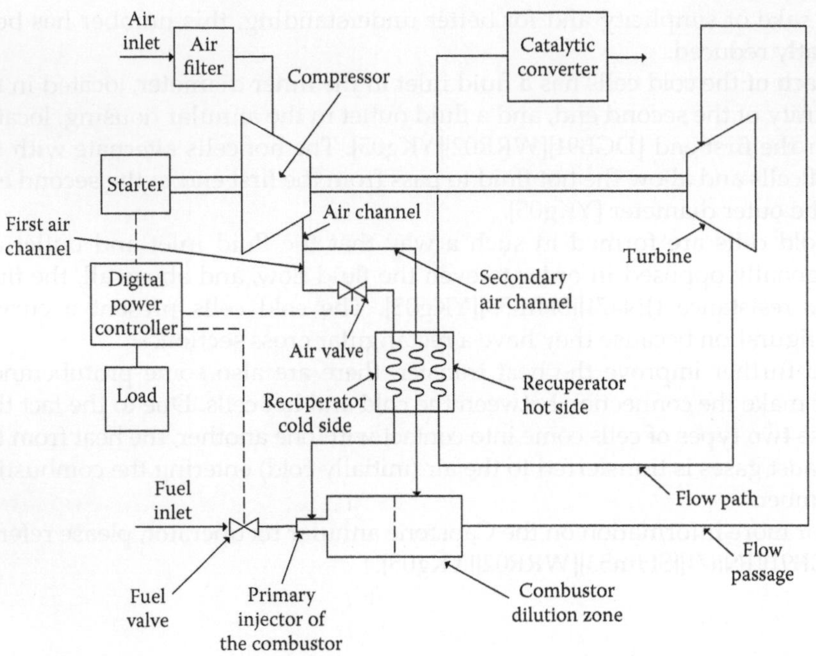


**FIGURE 2.12**
Approximate diagram of the Capstone low emissions natural gas MT. (From B. Dickey et al., Ultra low emissions gas turbine cycle using variable combustion primary zone airflow control. U.S. Patent 20020104316, issued August 8, 2002, available online: http://www.uspto.gov [accessed March 16, 2013]; and J. Delahaye et al., Gas turbine prime mover. U.S. Patent 3,798,898, issued March 26, 1974, available online: http://www.uspto.gov [accessed March 16, 2013]. With permission.)

the compressor through the first and secondary air channels. Ideally, the secondary air channel has to provide a constant air flow to the dilution passage and finally to the dilution zone [BDy02][JDe74]. This happens when the pressure upstream of the air valve is not influenced by the changes in the air flow passing through the valve [BDy02][JDe74]. Of course, these changes are directly influenced by how much the air valve is opened. In any case, the most important role of this air valve is to keep a constant AFR in the primary combustion zone of the combustor [BDy02].

The AFR is often determined through measurements for a certain level of power generation. Other AFRs can be determined for other corresponding power levels. Another key aspect is that the AFR has to have a value that can lead to a balanced minimization between $NO_x$ and CO [BDy02]. This is the most general case. The problem is much more complicated if we have, for example, a cluster of natural gas MTs that have to be operated in electrical load following mode and for which the minimization of $NO_x$ provokes the maximization of CO (and vice versa). That is because minimization of both

$NO_x$ and CO represents conflicting objectives. For further details about this, please refer to Chapter 3.

Coming back to the case of a single MT, the AFR also depends on the geometric properties of the combustor. Typical values for it which provide minimal emissions for $NO_x$ and CO are in the range of 30–40 for fuel flows of 0.002–0.025 $m^3/h$ [BDy02]. These values are valid for a combustor having an annular cross section, a flow path along the axis of maximum 30 cm, and a generated temperature of 1371°C–1649°C [BDy02].

The air resulting from the primary and dilution zones of the combustor is diverted through the turbine and then to the coiled pipes of the recuperator, and thus heat transfer from the exhaust gases to the air entering the combustion chamber takes place. The primary and the dilution zone passages in the recuperator (represented in Figure 2.12 by the first and second coiled pipes from left to right, respectively) present different air flows, and that is why AFR control is enabled in the primary zone [BDy02][JDe74].

The air exits the recuperator and enters first into a premix chamber where it is mixed with metered fuel provided by the fuel valve. The metered fuel is then mixed with primary zone air and finally burned. This process takes place based on the fact that the combustor permits the complete burning of the air–fuel mixture before it enters the dilution zone [BDy02]. Here, the burning products are further mixed with dilution air, and then directed to the turbine. The dilution air has two important features: it is already compressed and has thermal energy (obtained in the recuperator) [BDy02][JDe74]. In this way, the turbine begins to rotate and drives also the electrical generator. As can be seen in Figure 2.12, the turbine is coupled to a shaft (previously described) which makes the connection with the compressor and a starter motor.

At the next stage, the hot gases resulting from the turbine are diverted to the recuperator where these will increase the temperature of the air that enters the combustion chamber.

Coming back to the low emissions system, its core is represented by the air valve. The main role of this air valve is to maintain a constant AFR that will permit a balanced minimization of $NO_x$, CO, and other unwanted hydrocarbons. This is accomplished by keeping two separate air passages not only in the regenerator but also in the combustor. Given this, the air valve will be capable of forcing more air to enter the dilution zone, hence controlling the AFR [BDy02].

Theoretically, the air valve can be mounted either at the inlet or at the outlet of the recuperator [BDy02]. Coupling the valve at the inlet side would be the best due to the fact that it will be exposed to lower temperatures. This, of course, will result in a longer life and less maintenance costs.

Another important key component of the low emissions system implemented by Capstone is the catalytic reactor (see Figure 2.13). This represents the second stage of this system, after the air valve. This catalytic reactor works based on the oxidation principle of ingested fuel [CEJ80][DHn04][WCP75].

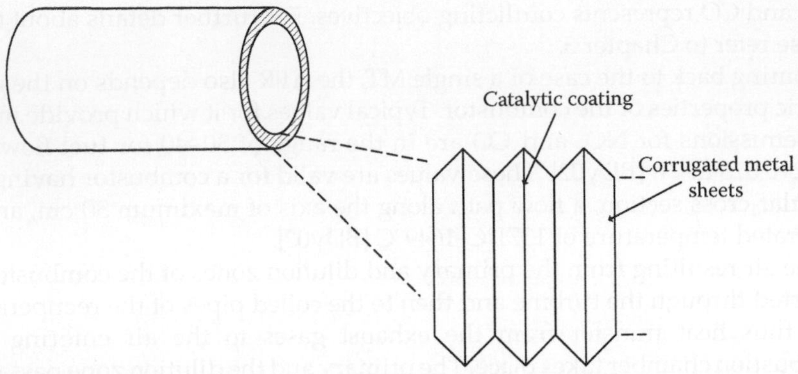


**FIGURE 2.13**
The Capstone catalytic reactor—a schematic diagram. (From D. Hamrin et al., Method for catalytic combustion in a gas-turbine engine, and applications thereof. U.S. Patent 20040148942, issued August 5, 2004, available online: http://www.uspto.gov [accessed March 16, 2013]. With permission.)

The exhaust gases from the catalytic reactor will actuate the turbine and thus the electrical generator.

In certain cases, the data related to the operation of the catalytic reactor is stored in special dedicated computer programs and further used in diagnostics [MBC93][DHn04]. The stored data could refer to [DHn04]:

- The temperature increase due to the changes in the fuel–air mixture being oxidized by the reactor
- Information about the unburned hydrocarbons
- Total operating time of the catalytic reactor

As shown in Figure 2.13, the catalytic reactor consists of a certain number of corrugated metal sheets which are coated with a special catalytic coating. This part of the MT has a large surface for oxidizing the air–fuel mixture [DHn04].

Given all these aspects, natural gas MTs, especially for single unit applications, have limitations [CEJ80][DHn04][MBC93]. One of the most important limitations is the fact that they cannot be operated on low-BTU (British Thermal Unit) fuels, and in this situation, undesirable pollutant emissions appear. Low BTU refers to the calorific value of the fuel used. For natural gas this is approximately 37.25 MJ/m$^3$, which is the equivalent of 999.759 BTU/scf (BTU/Standard Cubic Foot).

For more information about the pollutant emissions minimization system implemented by Capstone, please refer to [CEJ80][DHn04][MBC93][WCP75].

Different systems for pollutant emission minimization have been developed, but unfortunately they do not always work efficiently. In Chapter 3, an attempt is made to calculate and optimize through the evolutionary algorithm not only the pollutant emissions but also the fuel consumption for applications that require clusters of four MT either functioning in electrical load following mode or covering the load of a supermarket. In the second stage, the algorithm takes into account that the minimization of both $NO_x$ and CO represents conflicting objectives, and a compromise solution is obtained through Pareto front determination.

# 3

# *Gas Microturbines and Pollutant Emissions Optimization**

The latest trends in DG show that natural gas microturbines (MTs) are being increasingly adopted in urban areas (see Chapter 7, Case Studies) where air quality regulations are stringent. The pollution problem for a cluster of MTs has not received the attention it deserves until now, although all of the manufacturers produce highly reliable equipment from this point of view.

The problem of a cluster in which some of the MTs have to be shut down and others have to operate at full or partial load is complicated because many aspects have to be taken into consideration: the covering of the load, the fact that the experimental results show that when a MT is operated below 50% of its rated capacity the pollutant emissions tend to increase, or the fact that some of the components of these pollutant emissions have conflicting objectives (when one objective is minimized, the other is maximized and vice versa). The purpose of this chapter is to offer a solution to this problem based on the evolutionary algorithm (EA) and determination of the Pareto front.

Generally, the energy performance of an MT unit is expressed through its *electrical efficiency*:

$$\eta_W = W/F \tag{3.1}$$

where

$W$ = the electrical energy output [$kWh_e$]

$F$ = the fuel energy input [$kWh_t$]

Despite the fact that MTs can cogenerate heat and cooling with high overall efficiency [H97], only the problem of electrical energy production will be addressed, while thermal energy generation will be disregarded.

Electrical efficiency decreases when the MT is operated at partial load as a consequence of changes in thermodynamic cycle characteristics. Worsening of the combustion characteristics usually provokes an increase in pollutant emissions [CCM07], since some of these pollutants are generated by low flame in the combustion chamber (see Section 4.5). This problem is sometimes so severe that manufacturers advise consumers to switch their units off.

* Chapter 3 represents an improved version of [BCM08][BCM09][BCM11]. Extensive parts of this chapter have been republished or adapted from [BCM09] with the kind permission of the Institute of Electrical and Electronics Engineers (IEEE) and from [BCM11] with the kind permission of the Scientific Bulletin, University Politehnica of Bucharest.

The emission performance is characterized using an *emission factor model* [EDU-- ][EPA-- ]:

$$m^p = \mu^p \cdot W \tag{3.2}$$

where

$m^p$ = the mass emitted of a given pollutant $p$
$\mu^p$ = the emission factor (specific emissions) for a given pollutant $p$ [mg/kWh$_e$]
$W$ = the electrical energy output

The emission factor depends on various factors such as the operating conditions or the technology and size of the unit [EDU-- ][EPA-- ][Lo96] [PZR01].

## 3.1 Multi-Objective Optimization of Energy Efficiency and Pollutant Emissions

The following study is carried out given an hourly electrical load energy $W_{TOT}$ [kWh$_e$] that has to be supplied by a cluster of MT units operating in electrical load-following mode. This means that the MTs adjust their power output according to fluctuations in demand. Thus, the optimal share of the electrical load among the units in the cluster can be determined through the formulation of an appropriate optimization problem. The objectives to be minimized are:

- *$NO_x$ emissions,* which are the most hazardous pollutants for that equipment fed by natural gas [EPA-- ][Lo96], especially in urban areas. These types of emissions are usually subject to many stringent regulatory air quality constraints.
- *CO emissions,* on the other hand, are very low at full load and increase due to faulty maintenance, incomplete combustion at partial loads, or with aging of the components.
- *Fuel consumption* represents the energy efficiency goal. From an economical point of view, fuel consumption corresponds also to the minimization of costs incurred to purchase the fuel. Assuming that all MT units run on the same fuel (i.e., natural gas), fuel consumption minimization approximately corresponds to *$CO_2$ emission* minimization, according to the concepts discussed in [CCM07][CMN07]. The most important aspect here is that given a certain $W_{TOT}$, costs and $CO_2$ as well as energy efficiency are not supposed to be conflicting objectives under the hypotheses considered in this study.

Thus, this partitioning of the objectives is carried out under the assumption that $NO_x$ emissions, CO emissions, and fuel consumption are conflicting objectives, as discussed with numerical evidence below.

The constraints applied to this problem are decided by the operational limits of the equipment and the energy balance between MT generation and total load.

The analyses carried out in this chapter refer to the useful electrical output from the MTs. The *reference power* of each MT unit in the cluster is obtained by subtracting from the rated power an approximate amount of power needed to serve the auxiliary services of the unit (such as the gas compressor, in particular).

Let us consider a cluster of $i = 1,\ldots, N$ MTs, each of which has a reference power $P_i^{(r)}$ [$kW_e$]. The loading level $\alpha_i$ of the $i$-th MT unit in the cluster, for $i = 1, \ldots, N$, is expressed in relative units with respect to the reference power and varies within the range [0;1]. The minimum power of the $i$-th unit is symbolized by $P_i^{(\min)}$, while the constraint on the minimum loading of the MT unit is reflected on limiting the loading level within the range $[\alpha_i^{(\min)};1]$, whereas $\alpha_i^{(\min)} = P_i^{(\min)} / P_i^{(r)}$.

When operating at a certain loading level $\alpha_i$, the $i$-th MT unit is characterized by the specific $NO_x$ emissions $\mu_i^{NO_x}$ [$mg/kWh_e$], the specific CO emissions $\mu_i^{CO}$ [$mg/kWh_e$], and its electrical efficiency $\eta_i$, for $i = 1, \ldots, N$. Given a period $\tau = 1$ hour and an hourly energy $W_{TOT}$ supplied by the cluster of MTs to the load, the optimizations of the individual objectives are expressed as:

Minimization of overall $NO_x$ emissions:

$$\min \hat{f}^{NO_x}(W_{TOT}) = \sum_{i=1}^{N} \mu_i^{NO_x} \alpha_i P_i^{(r)} \tau \tag{3.3}$$

Minimization of overall CO emissions:

$$\min \hat{f}^{CO}(W_{TOT}) = \sum_{i=1}^{N} \mu_i^{CO} \alpha_i P_i^{(r)} \tau \tag{3.4}$$

Minimization of fuel (natural gas) consumption:

$$\min \hat{f}^{F}(W_{TOT}) = \sum_{i=1}^{N} \frac{\alpha_i P_i^{(r)} \tau}{\eta_i} \tag{3.5}$$

As explained, the constraints are given by the energy balance:

$$\sum_{i=1}^{N} \alpha_i P_i^{(r)} \tau - W_{TOT} = 0 \tag{3.6}$$

as well as by the loading level limits, for $i = 1, \ldots, N$:

$$\alpha_i \in \left\{ 0 \cup \left[ \alpha_i^{(\min)} ;1 \right] \right\} \tag{3.7}$$

For each objective $Z = \{NO_x, CO, F\}$, the formulation introduced previously is transformed into a penalized objective function, considering the penalty factor $\gamma$ applied to the energy balance constraint:

$$f^Z(W_{TOT}) = \hat{f}^Z(W_{TOT}) - \gamma \cdot \left| \sum_{i=1}^{N} \left( \alpha_i P_i^{(r)} \tau \right) - W_{TOT} \right| \tag{3.8}$$

s.t. (3.7).

The variables to be optimized are the loading levels $\eta_i$, for $i = 1, \ldots, N$. In addition to the non-connected loading level domain, the main challenges when computing the optimal solution depend on the non-linearity of the emission characteristics and energy efficiency, in the former case with possible non-monotonic emission profiles for variable MT loading. These non-monotonic emission characteristics generate a non-convex search space. Thus, the objective functions can be handled in two ways:

1. In the first case, the individual optimization problems can be solved separately. This process is carried out here using an EA, as described in [Go89]. The data of each MT unit (efficiency, emissions, and power) are coded using a discrete number of points, representing the switch-off condition and a predefined number of discrete loading levels within the range $[\alpha_i^{(\min)};1]$. More details on the EA formulation and application are illustrated below.
2. The interactions among various objectives can be addressed through a comprehensive multi-objective optimization framework. Given the presence of conflicting objectives, compromise solutions are obtained by determining the *Pareto front* which is composed of the *non-dominated* solutions for which it is not possible to improve the performance of one of the objectives without influencing another objective in a negative way. Another important aspect is that the identification of the entire Pareto front for multi-objective combinatorial optimization problems is practically infeasible [KCS06]. Thus, the purpose of the computational procedure has to be the obtaining of a *subset* of the Pareto front (also called *best-known Pareto front*) through direct calculation of a number of non-dominated solutions. The EA techniques are usually the most used and efficient for addressing this kind of problem [KCS06] [SD07].

In our particular case, the determination of the best-known Pareto front is made using a specific EA version adapted to solve multi-objective optimization problems. The specific characteristics of the EA, capable of operating with a population of solutions and of exploring different regions of the search space (in non-convex cases as well), are employed to obtain multiple non-dominated solutions in a single execution. More details on the EA version implemented here are provided in Section 3.2.

## 3.2 Multi-Objective Operational Optimization through the Evolutionary Algorithm

The two types of multi-objective optimization problems described in the previous section have been solved through specific EA programming tools. Given the circumstances of this particular case, these tools are different in their scope and implementation.

### 3.2.1 Individual Optimization

For solving the *individual optimization* problem, a classical EA version has been developed. The input data at partial load are the $NO_x$ and CO emission characteristics as well as the MT efficiencies. For information coding, the chromosome structure is formed by a number *N* of genes equal to the number of MTs. Each gene is characterized by *D* discrete states, each of which represents a specific operating level. Level 1 represents the switch-off condition. The other *D*-1 levels belong in the range $\left[\alpha_i^{(\min)};1\right]$, for $i = 1, \ldots, N$. In order to form the initial population of *K* chromosomes, some random levels are re-assigned to the genes.

At the same time, all the individual objective functions are positive valued. Due to the fact that the individual objectives must be minimized, while the EA intrinsically solves a maximization problem, each chromosome will be associated to a fitness (to be maximized) defined by using the inverse of the objective function; considering the *m*-th chromosome for the objective *Z*, its fitness will be:

$$\psi_m^Z = \frac{1/f_m^Z(W_{TOT})}{\sum_{v=1}^{M} 1/f_v^Z(W_{TOT})} \tag{3.9}$$

Further, the classical genetic operators (selection, crossover, and mutation) are applied in order to form a new population. The first of these operators, chromosome *selection*, is carried out through a mechanism called a "biased roulette wheel" in which the chromosomes of the new population are randomly selected on the basis of their fitness values. The next step is represented by the *crossover* which is applied to pairs of chromosomes of the selected population in case a random number extracted from a uniform probability distribution in the range [0;1] is lower than the user-defined crossover probability $p_C$; for those pairs of chromosomes satisfying the condition $r < p_C$, the crossover is carried out at a randomly chosen position. The *mutation*, on the other hand, is performed on a single gene, but its application is decided, like in the case of the *crossover*, through a two-step mechanism based on a user-defined mutation probability $p_M$ referred to a chromosome. Considering a random number $r$ extracted from a uniform probability distribution in the range [0;1], if for a chromosome the condition $r < p_M$ is satisfied, then a randomly chosen gene inside the chromosome has to suffer a mutation. The discrete loading level is represented by the fact that the gene is changed into a different loading level randomly chosen within the domain of definition of the $D$ loading levels mentioned above.

In our case, the *elitist* variant of the EA is adopted in the implementation, in which one copy of the chromosome corresponding to the best fitness is reproduced in the successive population without being modified by the selection, crossover, and mutation operators. This is done in order to obtain the best possible solution. The stop criterion is represented by the termination of the iterative process when there is no improvement in the best fitness over a predefined threshold $\varepsilon > 0$ for a specified number $H$ of successive iterations.

### 3.2.2 Pareto Front Construction

For a multi-objective problem with individual conflicting objectives, it is necessary to establish a sound criterion to manage the trade-off among the possible solutions. Given this, it is possible to exploit the concept of *Pareto dominance* [CN94]. Thus, a solution is considered to be *non-dominated* if none of the other solutions exhibits lower values of *all* the individual objective functions. On the other hand, the set of all the non-dominated points which can be defined for the multi-objective optimization problem forms the *Pareto front*. The solutions located on the Pareto front contain the optimal points corresponding to the application of the individual optimization criteria and also a number of *compromise solutions* in which none of the individual optima are attained.

The current literature indicates some methods suitable for performing *direct* construction of the *Pareto front*, or at least a portion of it, called the *best-known* Pareto front. Parts of these methods have reached an acceptably good effectiveness in simultaneously finding out a number of compromise solutions [DPA02][KCS06][SD07].

In our particular situation, the multi-optimization problem to be solved is strongly influenced by the presence of the equality constraint (3.6). Thus, a custom EA-based computational program with specific arrangements has been implemented for determining the best-known Pareto front, rather than using existing software. The most important difference compared to individual optimization is the definition and use of the fitness in the EA application. Regarding the Pareto front construction, higher fitness is assigned to the non-dominated points, and the genetic parameters in the EA are applied in such a way as to provide an appropriate variety of search in the space of the possible combinations of the discrete loading levels which can be represented in the genes of each chromosome.

Figure 3.1 presents the flowchart of the adapted EA algorithm. The general parameters are:

- Number $K$ of chromosomes
- Crossover probability $p_C$
- Mutation probability $p_M$
- Parameters $L$ and $\varepsilon$ used for the stop criteria
- Method-dependent parameter $\zeta$ which is used as a multiplier of the fitness values

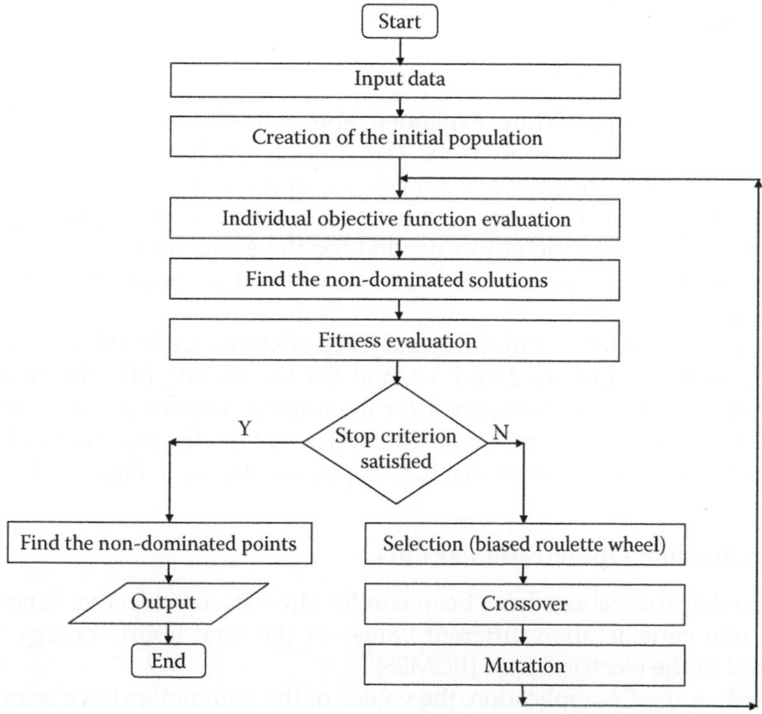


**FIGURE 3.1**
The flowchart of the EA-adapted procedure for determining the best-known Pareto front. (From A.V. Boicea, G. Chicco, and P. Mancarella, Optimal operation of a MT cluster with partial-load efficiency and emission characterization, Powertech, 2009 IEEE Bucharest, June 28–July 2, 2009, Pages 1–8. With permission.)

Chromosome coding for the $K$ chromosomes is the same as that used for the individual optimization addressed in Section 3.2. The initial population is created randomly. Fitness, on the other hand, is initialized at unity for all the chromosomes. The individual objective functions are then calculated. Given the objective functions, the non-dominated points are determined. At the next step, fitness is updated by multiplying by $\zeta$ the fitness of the chromosomes corresponding to the non-dominated solutions. Finally, global fitness is calculated as the sum of the fitness values.

The stop criterion is then tested and it is thus determined if no improvement in the global fitness over the threshold $\varepsilon$ has been detected after $L$ successive iterations. If the stop criterion is fulfilled, then the non-dominated points are found and the algorithm terminates with the output of the results. A new population is then created by applying the classical genetic operators of selection (through the biased roulette wheel described previously, on the basis of the fitness of each chromosome), crossover, and mutation.

## 3.3 Numerical Results

The optimization procedure explained above has been carried out on a cluster of four equal Capstone MTs. The chosen MTs have a rated power of 30 $kW_e$ and 60 $kW_e$, respectively. Of course, in the last case, due to the fact that the auxiliary installations also have to be driven, the reference power is 55 $kW_e$ (subtracting the power needed for the gas compressor operation, equal to about 5 $kW_e$ and assumed to be constant at partial load for the sake of simplicity).

The emission factors for the $NO_x$ and CO pollutants for 30 $kW_e$ and 60 $kW_e$ are indicated in Figures 3.2 and 3.3, and for the 60 $kW_e$ MT, the emission characteristics have been obtained for a sampled number of points elaborated from [PZR01] relevant to experimental results at discrete steps of 1 $kW_e$. The efficiency values for both machine types are shown in Figure 3.4.

### 3.3.1 Individual Optimization Results

Individual optimizations have been run for the various objective functions, taking into consideration different values of the total hourly energy $W_{TOT}$ delivered to the electrical load [BCM08].

Regarding the EA application, the values of the parameters have been chosen after a certain number of preliminary tests in order to balance solution effectiveness and computation time. The population is initially formed by $K = 100$ chromosomes, the crossover probability is $p_C = 0.6$, and the mutation probability $p_M = 0.1$. Compared to similar values from the current literature, this mutation probability is relatively high in order to allow more frequent

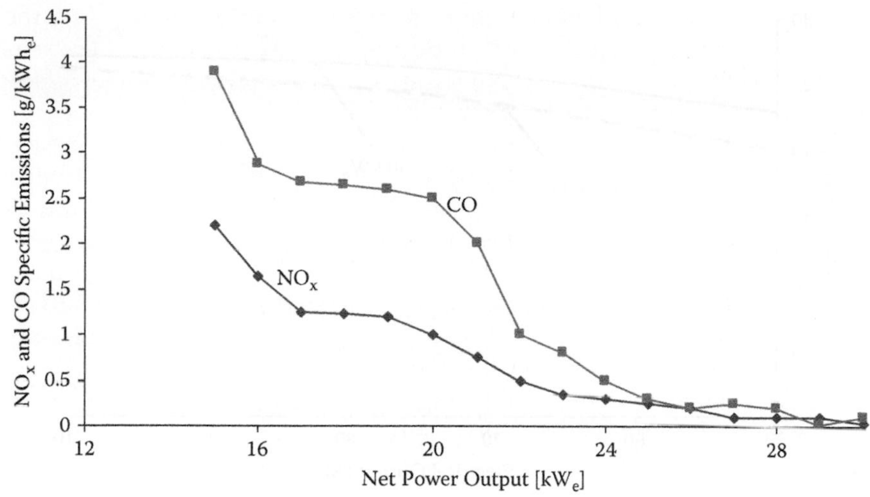


**FIGURE 3.2**
The $NO_x$ and CO emission characteristics for 30 kW$_e$ MT. (From A.V. Boicea, G. Chicco, and P. Mancarella, Optimal operation of a 30 kW natural gas MT cluster, *Buletinul Stiintific al Universitatii Politehnica Bucuresti, Seria C*, 73(1), 211–222, 2011. With permission.)

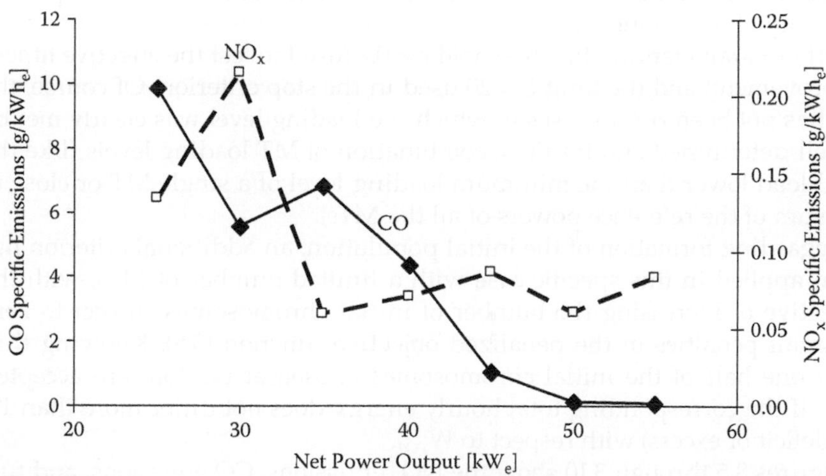


**FIGURE 3.3**
The $NO_x$ and CO emission characteristics for 60 kW$_e$ MT. (From A.V. Boicea, G. Chicco, and P. Mancarella, Optimal operation of a MT cluster with partial-load efficiency and emission characterization, Powertech, 2009 IEEE Bucharest, June 28–July 2, 2009, Pages 1–8. With permission.)

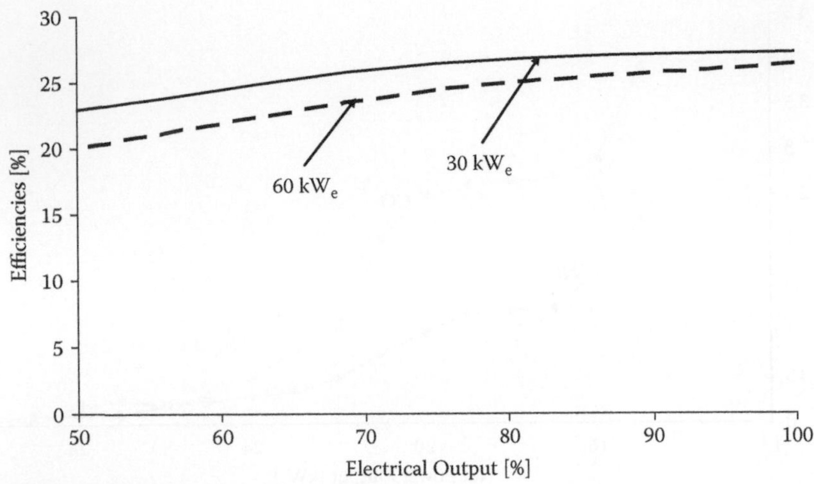


**FIGURE 3.4**
Electrical efficiency for 30 kW$_e$ and 60 kW$_e$ MT. (From A.V. Boicea, G. Chicco, and P. Mancarella, Optimal operation of a 30 kW natural gas MT cluster, *Buletinul Stiintific al Universitatii Politehnica Bucuresti, Seria C*, 73(1), 211–222, 2011. With permission.)

replacements of the discrete levels in the genes and thus an improved diversity of the solutions.

Other parameters are the threshold $\varepsilon = 0.1$ (used to test the effective fitness improvement) and the limit $L = 20$ used in the stop criterion. Of course, the EA has not been run in cases in which the loading level was clearly met by a well-determined and intuitive combination of MT loading levels (like the total load lower than the minimum loading level of a single MT or close to the sum of the reference powers of all the MTs).

Regarding formation of the initial population, an additional criterion has been applied in this specific case with a limited number of MTs, with the objective of increasing the number of initial chromosomes subject to zero or small penalties in the penalized objective function (3.8). Knowing this, only one half of the initial chromosomes chosen at random are accepted only if the corresponding total hourly energy does not differ more than 1% (in deficit or excess) with respect to $W_{TOT}$.

Figures 3.5 through 3.10 show the $NO_x$ emissions, CO emissions, and fuel consumption results for 30 kW$_e$ and 60 kW$_e$ MTs, respectively, obtained with the three optimization objectives. When comparing the optimal and non-optimal results, significant differences due to the effect of the opposite behavior of $NO_x$ and CO emissions in the intermediate partial-load operation region are obvious. Conversely, Figures 3.9 and 3.10 show no significant change in fuel consumption from the different optimization strategies.

The *usage* of the MT units at the various hourly energy values is presented in Figures 3.11 and 3.12 for minimum $NO_x$ emissions, in Figures 3.13 and 3.14

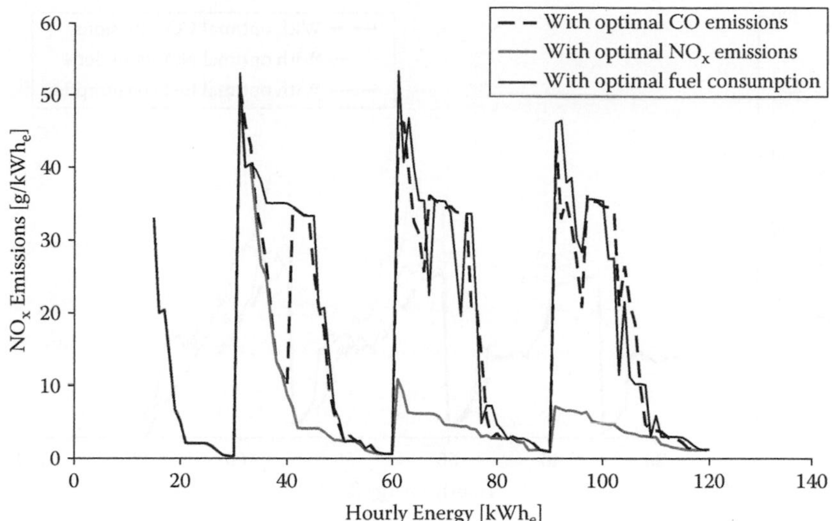


**FIGURE 3.5 (See color insert.)**
$NO_x$ emissions for different optimization scenarios in the case of a 30 kW$_e$ MT. (From A.V. Boicea, G. Chicco, and P. Mancarella, Optimal operation of a 30 kW natural gas MT cluster, *Buletinul Stiintific al Universitatii Politehnica Bucuresti, Seria C*, 73(1), 211–222, 2011. With permission.)

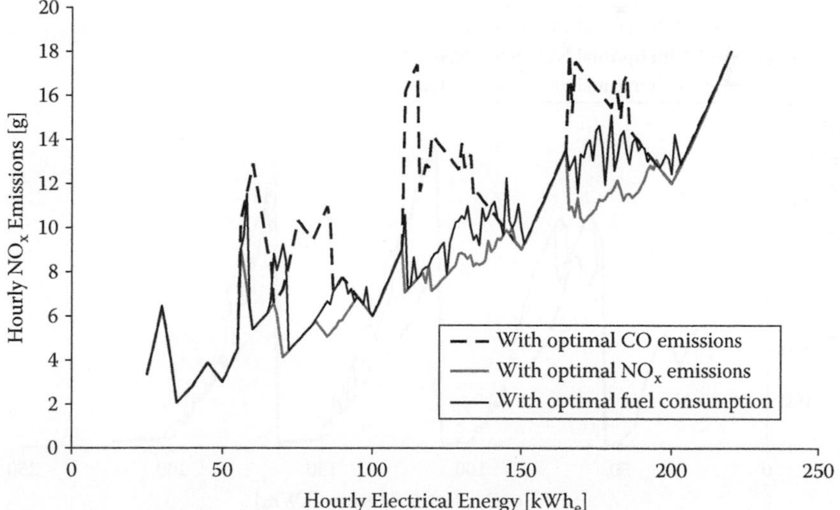


**FIGURE 3.6 (See color insert.)**
$NO_x$ emissions for different optimization scenarios in the case of a 60 kW$_e$ MT. (From A.V. Boicea, G. Chicco, and P. Mancarella, Optimal operation of a MT cluster with partial-load efficiency and emission characterization, Powertech, 2009 IEEE Bucharest, June 28–July 2, 2009, Pages 1–8. With permission.)

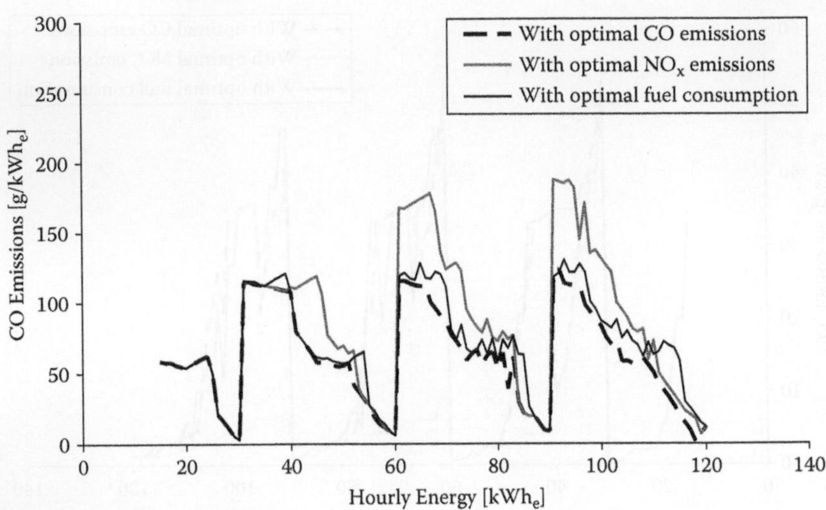


**FIGURE 3.7 (See color insert.)**
CO emissions for different optimization scenarios in the case of a 30 kW$_e$ MT. (From A.V. Boicea, G. Chicco, and P. Mancarella, Optimal operation of a 30 kW natural gas MT cluster, *Buletinul Stiintific al Universitatii Politehnica Bucuresti, Seria C*, 73(1), 211–222, 2011. With permission.)

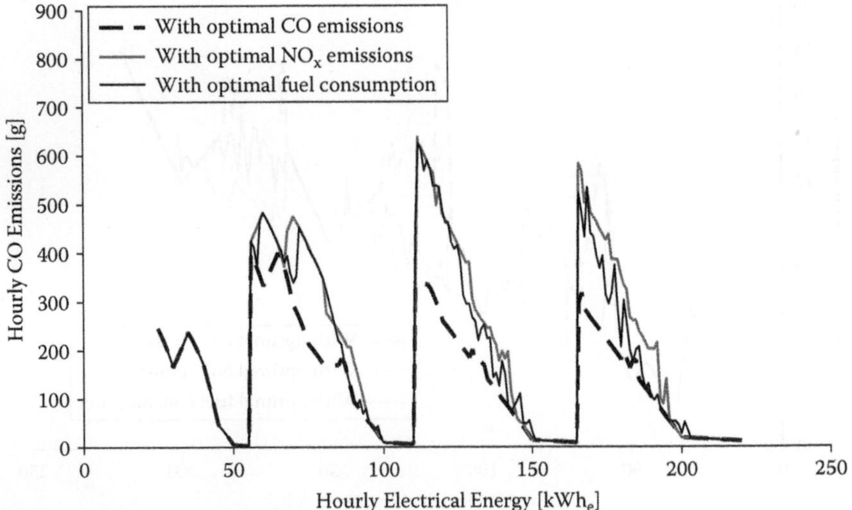


**FIGURE 3.8 (See color insert.)**
CO emissions for different optimization objectives in the case of a 60 kW$_e$ MT. (From A.V. Boicea, G. Chicco, and P. Mancarella, Optimal operation of a MT cluster with partial-load efficiency and emission characterization, Powertech, 2009 IEEE Bucharest, June 28–July 2, 2009, Pages 1–8. With permission.)

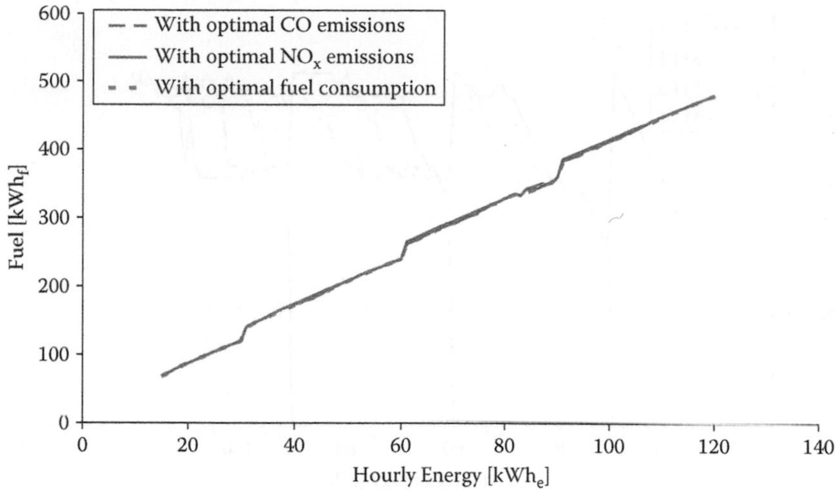


**FIGURE 3.9 (See color insert.)**
Fuel consumption for different optimization scenarios in the case of a 30 $kW_e$ MT. (From A.V. Boicea, G. Chicco, and P. Mancarella, Optimal operation of a 30 kW natural gas MT cluster, *Buletinul Stiintific al Universitatii Politehnica Bucuresti, Seria C*, 73(1), 211–222, 2011. With permission.)

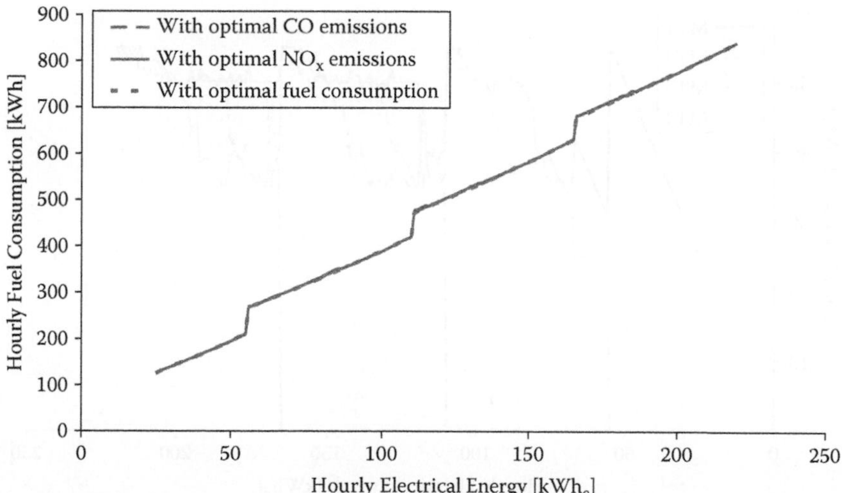


**FIGURE 3.10 (See color insert.)**
Fuel consumption for different optimization objectives in the case of a 60 $kW_e$ MT. (From A.V. Boicea, G. Chicco, and P. Mancarella, Optimal operation of a MT cluster with partial-load efficiency and emission characterization, Powertech, 2009 IEEE Bucharest, June 28–July 2, 2009, Pages 1–8. With permission.)

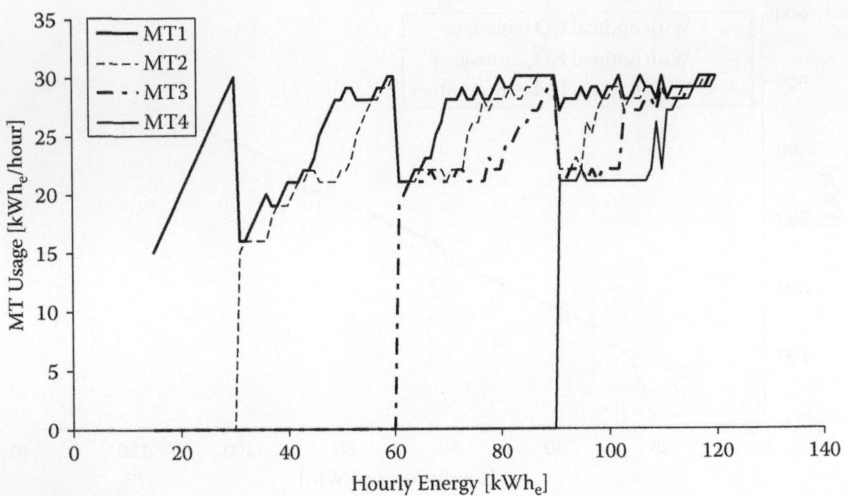


**FIGURE 3.11**
MT usage for optimal $NO_x$ emissions in the case of a 30 kW$_e$ MT. (From A.V. Boicea, G. Chicco, and P. Mancarella, Optimal operation of a 30 kW natural gas MT cluster, *Buletinul Stiintific al Universitatii Politehnica Bucuresti, Seria C*, 73(1), 211–222, 2011. With permission.)

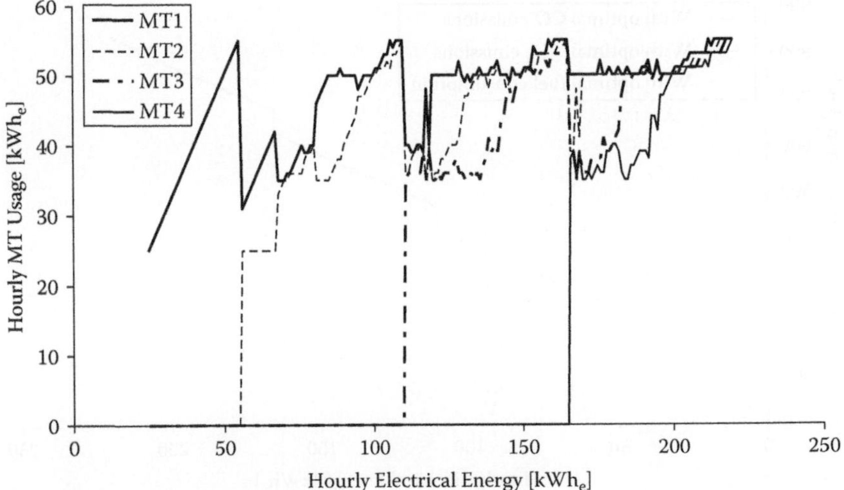


**FIGURE 3.12**
MT usage for optimal $NO_x$ emissions in the case of a 60 kW$_e$ MT. (From A.V. Boicea, G. Chicco, and P. Mancarella, Optimal operation of a MT cluster with partial-load efficiency and emission characterization, Powertech, 2009 IEEE Bucharest, June 28–July 2, 2009, Pages 1–8. With permission.)

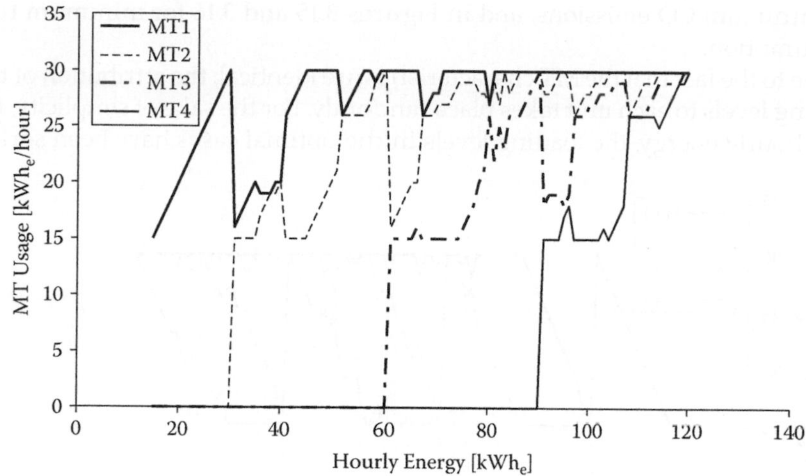


**FIGURE 3.13**
MT usage for optimal CO emissions in the case of a 30 $kW_e$ MT. (From A.V. Boicea, G. Chicco, and P. Mancarella, Optimal operation of a 30 kW natural gas MT cluster, *Buletinul Stiintific al Universitatii Politehnica Bucuresti, Seria C*, 73(1), 211–222, 2011. With permission.)

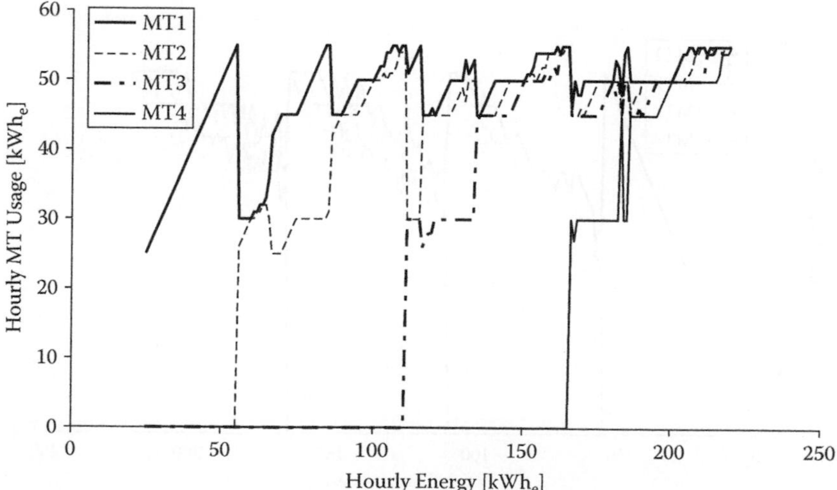


**FIGURE 3.14**
MT usage for optimal CO emissions in the case of a 60 $kW_e$ MT. (From A.V. Boicea, G. Chicco, and P. Mancarella, Optimal operation of a MT cluster with partial-load efficiency and emission characterization, Powertech, 2009 IEEE Bucharest, June 28–July 2, 2009, Pages 1–8. With permission.)

for minimum CO emissions, and in Figures 3.15 and 3.16 for minimum fuel consumption.

Due to the fact that the MT characteristics are identical, the attribution of the loading levels to each unit takes place randomly. For the sake of simplicity, for each hourly energy, the loading levels in the optimal cases have been sorted

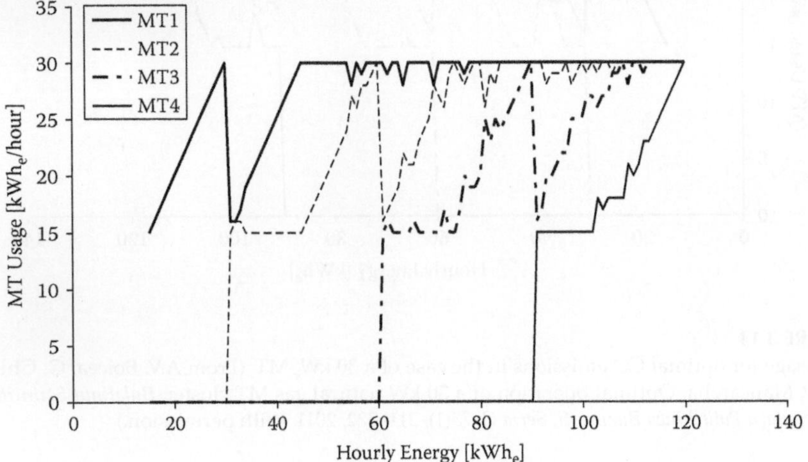


**FIGURE 3.15**
MT usage for optimal fuel consumption in the case of a 30 kW$_e$ MT.

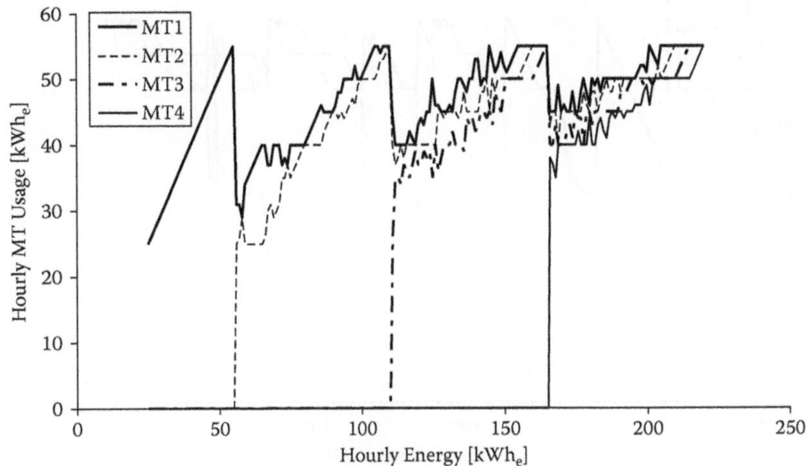


**FIGURE 3.16**
MT usage with optimal fuel consumption for a 60 kW$_e$ MT. (From A.V. Boicea, G. Chicco, and P. Mancarella, Optimal operation of a MT cluster with partial-load efficiency and emission characterization, Powertech, 2009 IEEE Bucharest, June 28–July 2, 2009, Pages 1–8. With permission.)

in descending order, assigning the highest loading level to unit MT1, the successive value in descending order to unit MT2, and so forth. In reality, the operational schedule of the units has to be analyzed considering specific load patterns in the time domain (as in the next section) and taking into account further operational limits (for instance, the number of switch-on/-off operations during the day, in order to avoid maintenance problems).

### 3.3.2 Application to a Commercial Center

The individual optimization results obtained above are applied to an illustrative real case—a commercial center. A similar situation is presented in [BCM08].

The electrical load pattern data of this supermarket, containing the total electricity consumption, has been measured in the field and is represented in sampled form at discrete time steps of 15 minutes each (Figure 3.17). The electrical demand of the commercial center is covered using the cluster of $N = 4$ MTs of 60 $kW_e$.

As observed, different patterns of electricity generation from the MTs are formed at each point in time on the basis of the results found from the optimizations carried out by applying the different individual objectives. Figure 3.18 presents comparisons between the 15-min $NO_x$ emissions results. One can see that the $NO_x$ emission pattern corresponding to the optimal CO results exhibits a large increase with respect to that representing optimal $NO_x$ emissions, while the $NO_x$ emission worsening obtained by using the minimum fuel consumption strategy is lower. This depends on the different

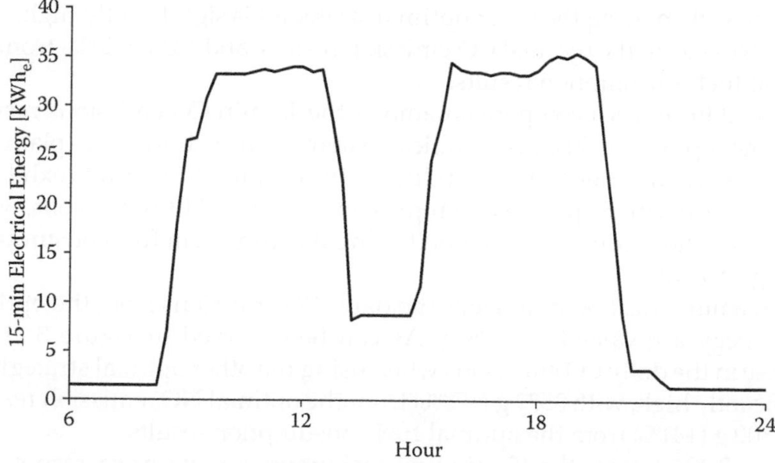


**FIGURE 3.17**
Daily electrical load profile of the supermarket. (From A.V. Boicea, G. Chicco, and P. Mancarella, Optimal operation of a MT cluster with partial-load efficiency and emission characterization, Powertech, 2009 IEEE Bucharest, June 28–July 2, 2009, Pages 1–8. With permission.)

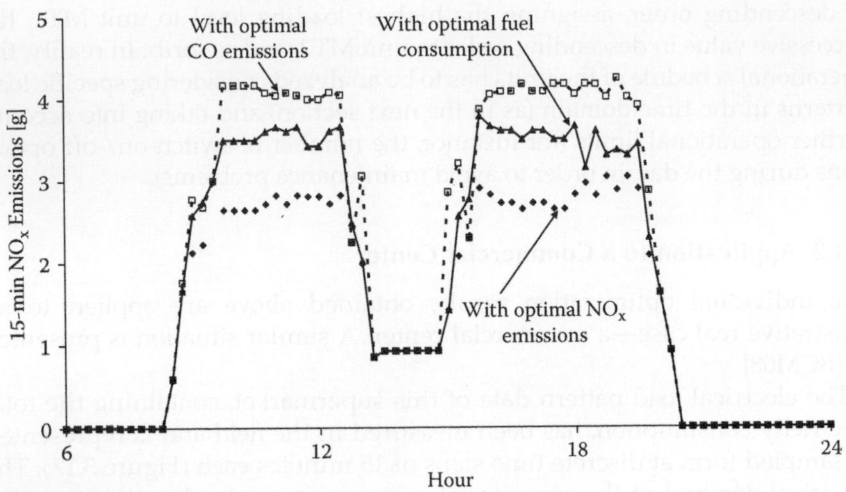


**FIGURE 3.18**
$NO_x$ emissions for the three optimization contexts in the case of a 60 $kW_e$ MT. (From A.V. Boicea, G. Chicco, and P. Mancarella, Optimal operation of a MT cluster with partial-load efficiency and emission characterization, Powertech, 2009 IEEE Bucharest, June 28–July 2, 2009, Pages 1–8. With permission.)

trends of $NO_x$ and CO emissions, as mentioned. With this example it is possible to give an appreciation of this worsening in a real situation. It has been calculated that the daily $NO_x$ emissions from the optimal $NO_x$ strategy are valued at 109 g. On the other hand, the increase in the daily $NO_x$ emissions when using the other optimal strategies is significantly high, with 151 g (+39%) from the optimal CO emission results, and 132 g (+21%) from the optimal fuel consumption results.

Figure 3.19 shows a comparison among the 15-min CO emissions results. The same type of conclusions as indicated above can be drawn to explain that the CO emission pattern corresponding to the optimal $NO_x$ results exhibits a large increase with respect to that representing optimal CO emissions, while the CO emission worsening obtained using the minimum fuel consumption strategy is lower.

From a numerical point of view, the daily CO emissions from the optimal CO strategy are valued at 1986 g. As can be observed in Figure 3.19, the increase in the daily CO emissions when using the other optimal strategies is significantly high, with 3532 g (+78%) from the optimal $NO_x$ emission results, and 2800 g (+41%) from the optimal fuel consumption results.

Figure 3.20 contains the 15-min fuel consumption patterns corresponding to the three optimization criteria. In this case the results overlap, and almost no significant difference is visible.

Other possible *usage* patterns for the individual MT units are shown for the minimum $NO_x$ emission case, in Figure 3.21 for the minimum CO emission

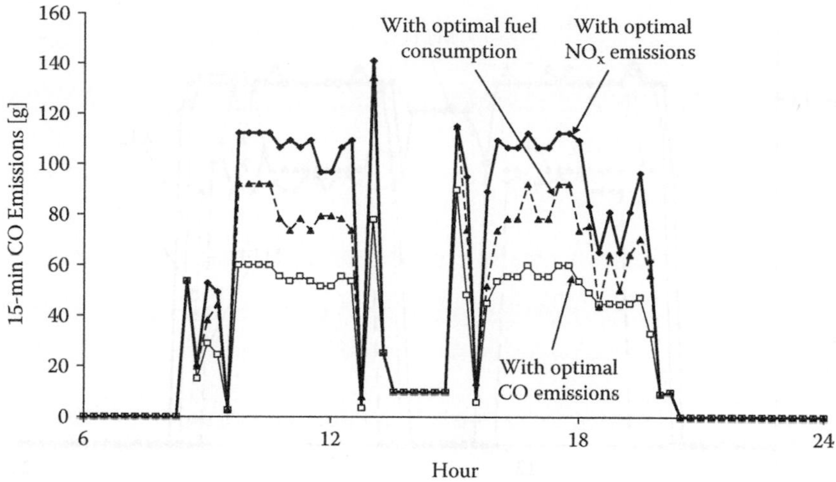


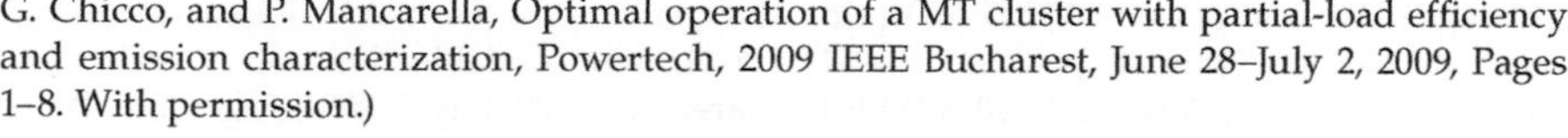

**FIGURE 3.19**
CO emissions for the three optimization contexts in the case of a 60 $kW_e$ MT. (From A.V. Boicea, G. Chicco, and P. Mancarella, Optimal operation of a MT cluster with partial-load efficiency and emission characterization, Powertech, 2009 IEEE Bucharest, June 28–July 2, 2009, Pages 1–8. With permission.)

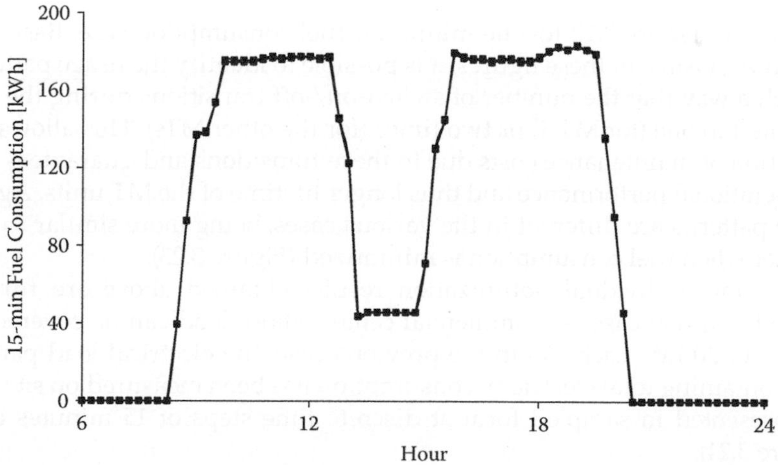


**FIGURE 3.20**
Fuel consumption in the case of a 60 $kW_e$ MT for the three optimization contexts (results overlap). (From A.V. Boicea, G. Chicco, and P. Mancarella, Optimal operation of a MT cluster with partial-load efficiency and emission characterization, Powertech, 2009 IEEE Bucharest, June 28–July 2, 2009, Pages 1–8. With permission.)

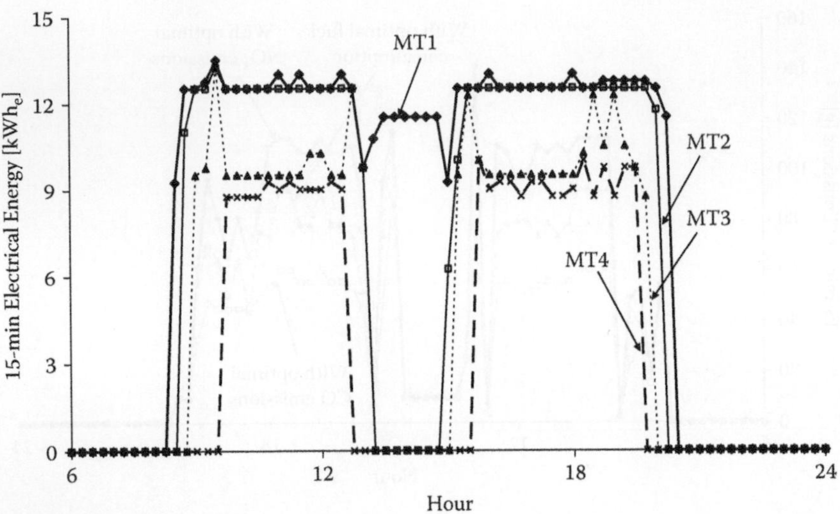


**FIGURE 3.21**
MT usage in the case of a 60 kW$_e$ MT for optimal $NO_x$ emissions. (From A.V. Boicea, G. Chicco, and P. Mancarella, Optimal operation of a MT cluster with partial-load efficiency and emission characterization, Powertech, 2009 IEEE Bucharest, June 28–July 2, 2009, Pages 1–8. With permission.)

case and in Figure 3.22 for the minimum fuel consumption case. Based on the curves visible in these figures, it is possible to identify the usage patterns in such a way that the number of switch-on/-off transitions during the day is limited to one (for MT 1) or two times (for the other MTs). This allows the reduction of maintenance costs due to these transitions and guarantees better operational performance and thus longer lifetime of the MT units. Again, usage patterns are different in the various cases, being more similar to one another when fuel consumption is minimized (Figure 3.23).

The same individual optimization results obtained above are further applied to a real case—a commercial center whose load can be covered by 4 units of 30 kW$_e$ each. As in the previous case, the electrical load pattern data containing total electricity consumption has been measured on site and is represented in sampled form at discrete time steps of 15 minutes each (Figure 3.24).

Figure 3.25 presents comparisons among the 15-min $NO_x$ emissions results. As can be observed, the $NO_x$ emission pattern corresponding to the optimal CO results exhibits an important increase compared to the one representing the optimal $NO_x$ emissions, while the $NO_x$ emission worsening obtained by using the minimum fuel consumption strategy is lower. This again, as in the previous case, depends on the different trends of $NO_x$ and CO emissions. Clearly, with this example it is possible to give a quantification of this worsening in a real situation.

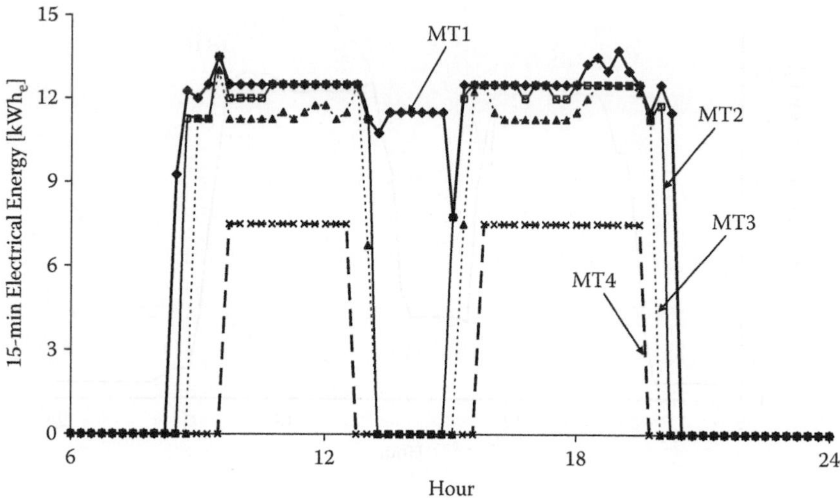


**FIGURE 3.22**
MT usage in the case of a 60 kW$_e$ MT for optimal CO emissions. (From A.V. Boicea, G. Chicco, and P. Mancarella, Optimal operation of a MT cluster with partial-load efficiency and emission characterization, Powertech, 2009 IEEE Bucharest, June 28–July 2, 2009, Pages 1–8. With permission.)

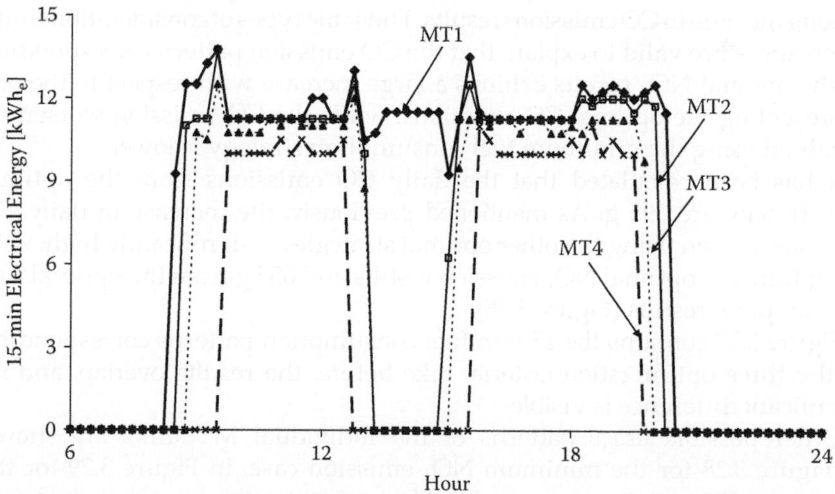


**FIGURE 3.23**
MT usage in the case of a 60 kW$_e$ MT for optimal fuel consumption. (From A.V. Boicea, G. Chicco, and P. Mancarella, Optimal operation of a MT cluster with partial-load efficiency and emission characterization, Powertech, 2009 IEEE Bucharest, June 28–July 2, 2009, Pages 1–8. With permission.)

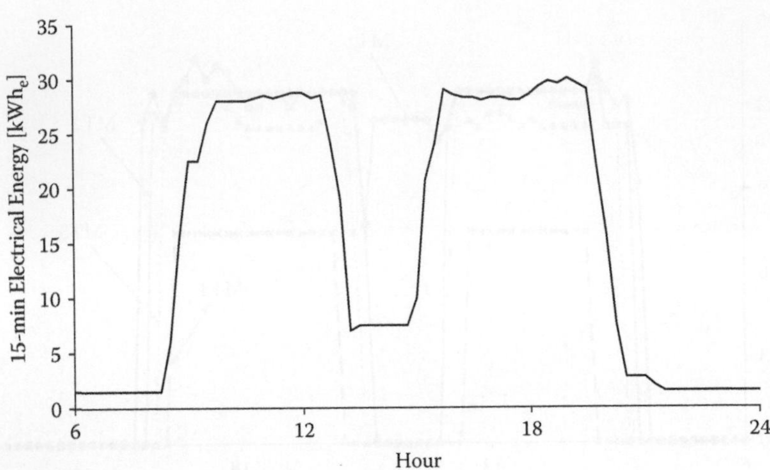


**FIGURE 3.24**
The daily electrical load profile of the supermarket.

It has been calculated that daily $NO_x$ emissions from the optimal $NO_x$ strategy are 25 g. The increase in the daily $NO_x$ emissions when using the other optimal strategies is significantly high, with 69 g from the optimal CO emission results and 82 g from the optimal fuel consumption results.

Figure 3.25 shows the comparison given the three optimization scenarios among the 15-min CO emissions results. The same types of considerations indicated above are valid to explain that the CO emission pattern corresponding to the optimal $NO_x$ results exhibits a large increase with respect to the one representing the optimal CO emissions, while the CO emission worsening obtained using the minimum fuel consumption strategy is lower.

It has been calculated that the daily CO emissions from the optimal CO strategy are 395 g. As mentioned previously, the increase in daily CO emissions when using the other optimal strategies is significantly high, with 520 g from the optimal $NO_x$ emission results and 635 g from the optimal fuel consumption results (Figure 3.26).

Figure 3.27 contains the 15-min fuel consumption patterns corresponding to the three optimization criteria. Like before, the results overlap, and no significant difference is visible.

Other possible *usage* patterns of the individual MT units are shown in Figure 3.28 for the minimum $NO_x$ emission case, in Figure 3.29 for the minimum CO emission case, and in Figure 3.30 for the minimum fuel consumption case. Based on the curves presented in these figures, it is possible to identify in this case the usage patterns in such a way that the number of switch-on/-off transitions during the day is limited to one (for MT 1) or two times (for the other MTs).

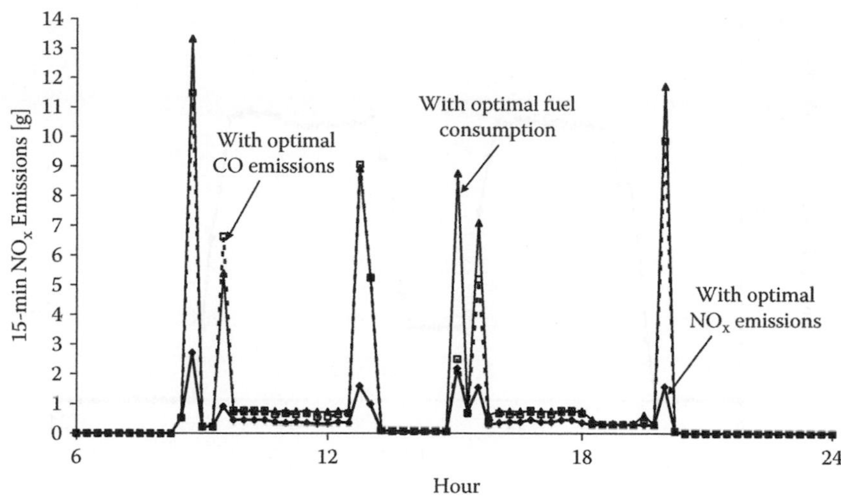


**FIGURE 3.25**
$NO_x$ emissions for the three optimization contexts in the case of a 30 $kW_e$ MT.

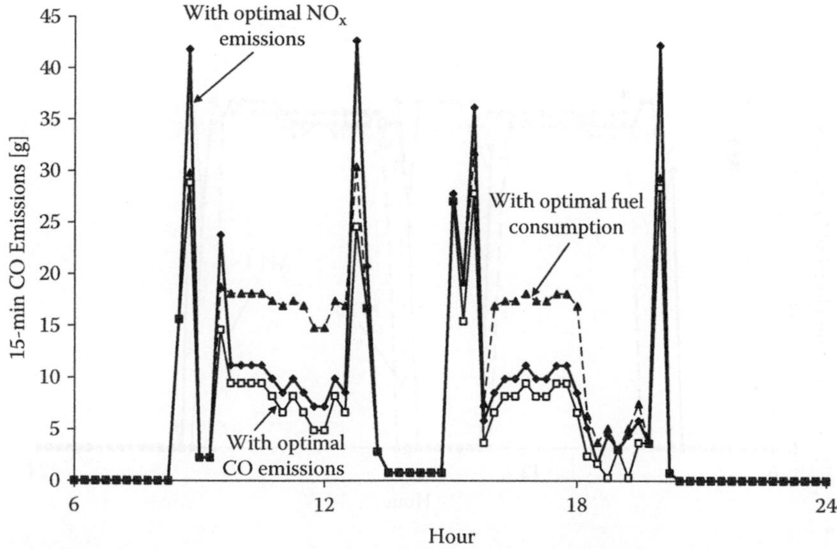


**FIGURE 3.26**
CO emissions for the three optimization contexts in the case of a 30 $kW_e$ MT.

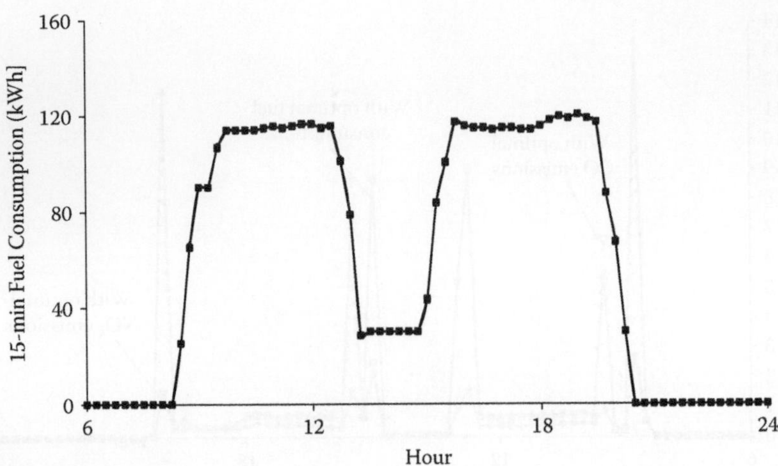


**FIGURE 3.27**
Fuel consumption for a 30 $kW_e$ MT for the three optimization contexts (results overlap).

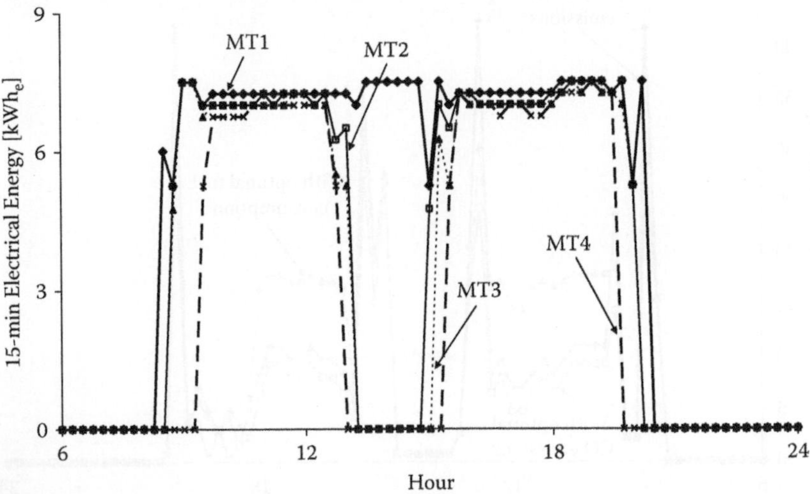


**FIGURE 3.28**
MT usage in the case of a 30 $kW_e$ MT for optimal $NO_x$ emissions scenario.

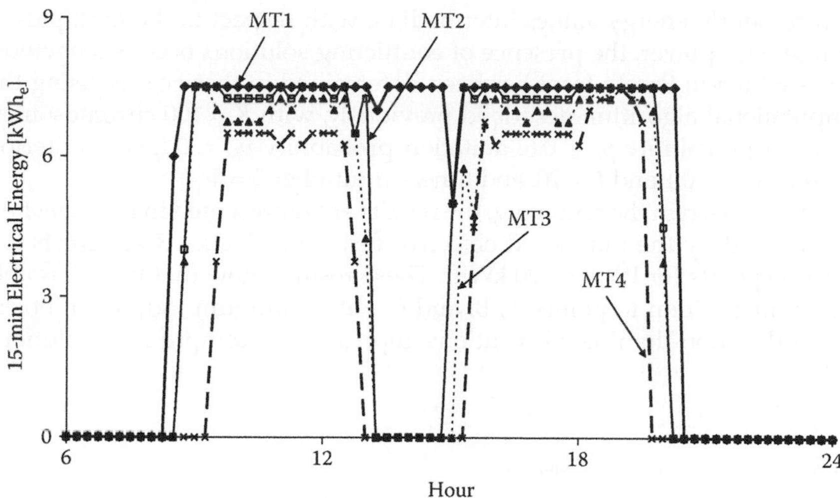


**FIGURE 3.29**
MT usage in the case of a 30 kW$_e$ MT for optimal CO emissions scenario.

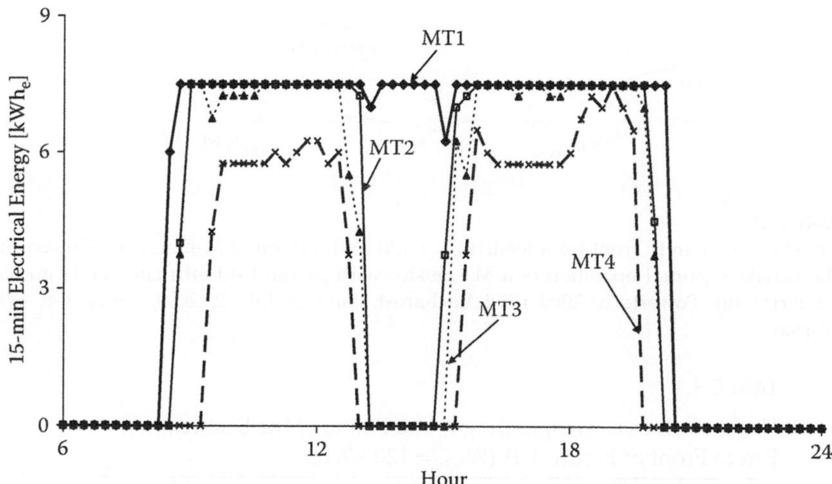


**FIGURE 3.30**
MT usage in the case of a 30 kW$_e$ MT for optimal fuel consumption scenario.

### 3.3.3 Pareto Analysis Results

All the results obtained from the individual objectives optimization are helpful to observe whether conflicting results arise from adopting the different objective functions. For our specific case with equal MT units, the solutions clearly tend to be similar when the hourly energy is slightly lower or equal to a multiple of the reference power.

Given hourly energy values intermediate with respect to the multiples of the reference power, the presence of conflicting solutions becomes obvious. The best-known Pareto front has been determined in these cases, using the computational algorithm described previously, with $K = 100$ chromosomes, crossover probability $p_C = 0.6$, mutation probability $p_M = 0.1$, stop criterion parameters $\varepsilon = 0.1$ and $L = 20$, and fitness multiplier $\zeta = 4$.

As can be seen in the following, the results are represented in the *phenotype space* formed by the individual objective functions. Figure 3.31 depicts the resulting points for $W_{TOT} = 120$ kWh$_e$. These results show that the EA is able to find, in addition to points A, B, and C with minimum individual objectives, other non-dominated solutions representing compromise solutions

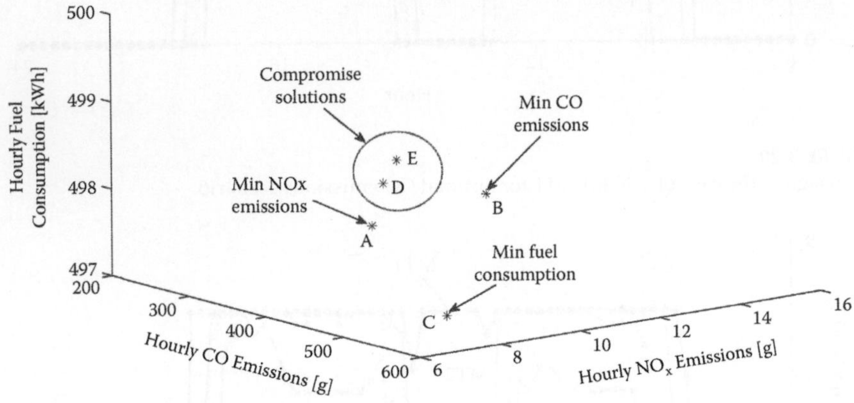


**FIGURE 3.31**
The best-known Pareto front for a load $W_{TOT} = 120$ kWh$_e$. (From A.V. Boicea, G. Chicco, and P. Mancarella, Optimal operation of a MT cluster with partial-load efficiency and emission characterization, Powertech, 2009 IEEE Bucharest, June 28–July 2, 2009, Pages 1–8. With permission.)

**TABLE 3.1**
Hourly Values Corresponding to the Points of the Best-Known Pareto Front of Figure 3.31 ($W_{TOT} = 120$ kWh$_e$)

| Point | MT 1 (kWh$_e$) | MT 2 (kWh$_e$) | MT 3 (kWh$_e$) | MT 4 (kWh$_e$) | $NO_x$ (g) | CO (g) | Fuel (kWh) |
|---|---|---|---|---|---|---|---|
| A | 50 | 35 | 35 | 0 | **7.13** | 479.3 | 498.1 |
| B | 45 | 45 | 30 | 0 | 14.19 | **259.3** | 497.5 |
| C | 42 | 40 | 38 | 0 | 8.54 | 501.5 | **497.1** |
| D | 49 | 36 | 35 | 0 | 7.45 | 477.6 | 498.6 |
| E | 50 | 36 | 34 | 0 | 8.26 | 453.2 | 498.8 |

*Source:* A.V. Boicea, G. Chicco, and P. Mancarella, Optimal operation of a MT cluster with partial-load efficiency and emission characterization, Powertech, 2009 IEEE Bucharest, June 28–July 2, 2009, Pages 1–8. With permission.

for the multi-objective optimization problem. The numerical results are also reported in Table 3.1 in order to check numerically that they correspond to non-dominated solutions. The results presented in Figure 3.31 and Table 3.1 refer to a single value of the total load.

For a broader picture, the best-known Pareto points are again depicted for values of $W_{TOT}$ variable in the range 110–220 $kW_e$, corresponding for example to the range of interest for the MT cluster operation at a relatively high total load in the commercial center case study. The representation has been split into three figures, and only the case of the 60kW MT has been chosen, since the two situations are similar. This has been done to give the whole picture of the components of the objective array $Z$ through three-dimensional views. The common part of these figures is the hourly energy axis. On the other axes, Figure 3.32 describes the hourly CO emissions and the hourly fuel consumption, Figure 3.33 reports the hourly $NO_x$ emissions and the hourly fuel consumption, and Figure 3.34 shows the hourly $NO_x$ emissions and the hourly CO emissions. These figures give an interesting perspective of the impact of the objective functions (in particular the different behavior of the specific $NO_x$ and CO emissions at partial loads) on providing different optima and compromise solutions.

The graphical representations with the best-known Pareto front in the case of a 30 kW MT are similar to those corresponding to the 60 kW MT. For this reason, they have not been included in this work.

Observing all these results, one can say that these optimization strategies do not greatly influence fuel consumption but are important for pollutant emissions. On the other hand, MT usage patterns show that operation in almost optimal conditions is possible without requiring an excessive number of switch-on/-off operations during one day. This is important for the life of the MT and also for the maintenance costs.

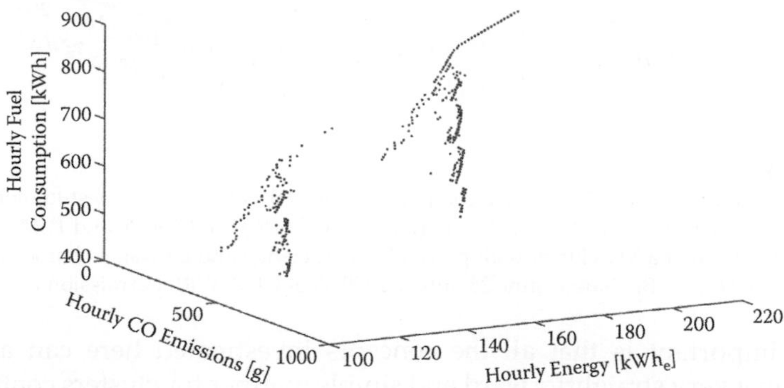


**FIGURE 3.32**
Hourly fuel consumption and CO emission components for a 60 $kW_e$ MT of the best-known Pareto front for $W_{TOT}$ variable from 110–220 $kWh_e$. (From A.V. Boicea, G. Chicco, and P. Mancarella, Optimal operation of a MT cluster with partial-load efficiency and emission characterization, Powertech, 2009 IEEE Bucharest, June 28–July 2, 2009, Pages 1–8. With permission.)

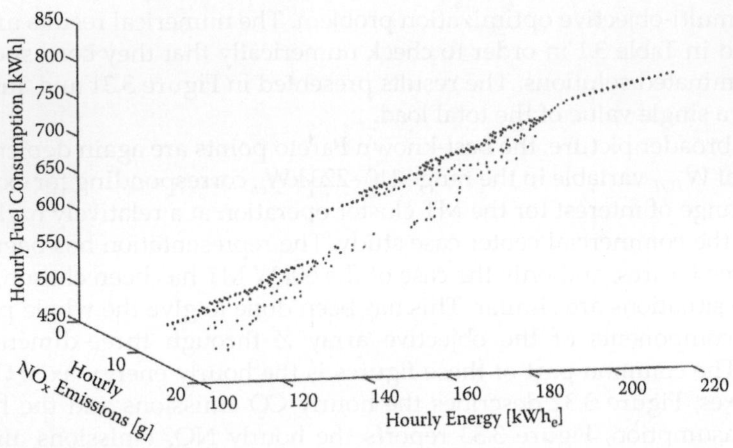


**FIGURE 3.33**
Hourly fuel consumption and $NO_x$ emission components for a 60 kW$_e$ MT of the best-known Pareto front for $W_{TOT}$ variable from 110–220 kWh$_e$. (From A.V. Boicea, G. Chicco, and P. Mancarella, Optimal operation of a MT cluster with partial-load efficiency and emission characterization, Powertech, 2009 IEEE Bucharest, June 28–July 2, 2009, Pages 1–8. With permission.)

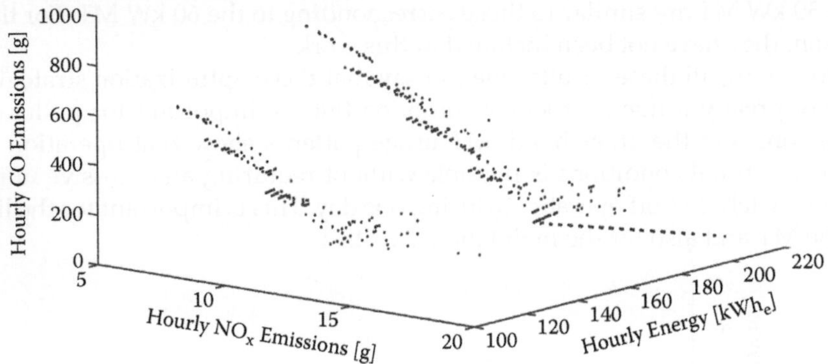


**FIGURE 3.34**
Hourly CO emission and $NO_x$ emission components for a 60 kW$_e$ MT of the best-known Pareto front for $W_{TOT}$ variable from 110–220 kWh$_e$. (From A.V. Boicea, G. Chicco, and P. Mancarella, Optimal operation of a MT cluster with partial-load efficiency and emission characterization, Powertech, 2009 IEEE Bucharest, June 28–July 2, 2009, Pages 1–8. With permission.)

More important is that all the concepts investigated here can also be applied in a very straightforward and simple manner for clusters containing more than four units or containing units having different rated powers.

With this chapter the detailed description of the pollutant emissions generated by natural gas MTs comes to an end. On the following pages, more emphasis is placed on their electrical performances. Another type of MT is chosen, and its design is thoroughly described in Chapter 4.

# 4

# Generalities on the Design of a TA-100 Natural Gas Microturbine

The Elliott TA-100 is a natural gas microturbine (MT) with a rated power of 100 kW. It was initially manufactured by Elliott Energy Systems and at present is manufactured by Capstone.

## 4.1 The Gas Compressor

As with the previously described Capstone MTs, the Elliott turbogenerator needs compressed gas due to the fact that in the majority of locations (especially urban areas) in which this equipment is deployed, gas pressure is very low, being valued at approximately 0.2–0.3 psig (see Chapter 2).

In the 1990s, most MT applications used reciprocating compressors, but with the inventions of Capstone and Elliott the situation changed for the better. Two of the most important disadvantages of the reciprocating compressors are that they induce undesirable vibrations to the whole system and they are expensive to operate and maintain [HHa02][SVv05] [YHa97].

Compared to the Capstone turbogenerator, some of the Elliott systems make use of a separate air compressor, as shown in Figure 4.1 [SVv05]. In our particular case, the MT on which the measurements in Chapter 6 have been carried out does not have such an air compressor.

The gas compressor is radial and is powered by an electrical motor which during the starting sequence of the MT is fed with electrical energy from a battery. Another important element that was characteristic of the Elliott system was the "air bearing compressor," whose role was to provide compressed air to the air bearings of the MT. In the moment when the starting sequence is completed, the power turbine will power both the gas and the air compressors [PGL64][SVv05].

Depending on the application, the turbogenerator can have either one or two shafts (described in Figure 4.1). Most of the applications use a single shaft due to the compactness and reliability [HHa02][SVv05][YHa97].

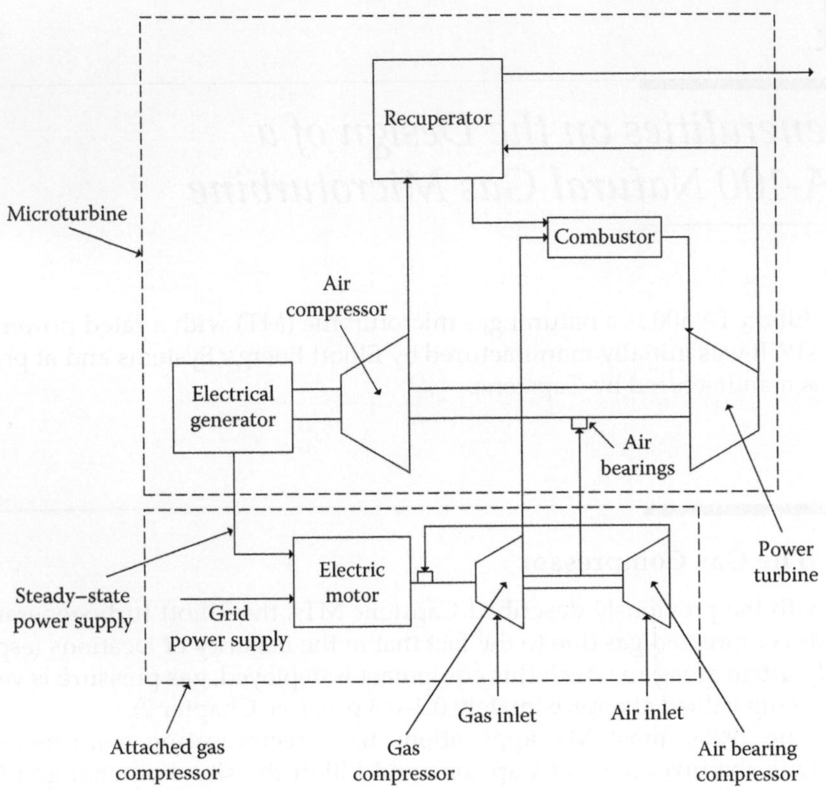


**FIGURE 4.1**
The Elliott gas compressor system. (From D. Jensen, MT power generating system including variable-speed gas compressor. U.S. Patent 6,066,898, issued May 23, 2000, available online: http://www.uspto.gov [accessed March 17, 2013]; and S. Voinov, Gas compression system and method for MT application. U.S. Patent 6,892,542, issued May 17, 2005, available online: http://www.uspto.gov [accessed March 17, 2013]. With permission.)

The electric starter motor is designed in such a way that it brings the air compressor to a predetermined speed in order to supply the appropriate quantity of air needed by the above-mentioned air bearings [PGL64][SVv05].

For further information on the Elliott gas compressor, please refer to [DJn00][HHa02][PGL64][SVv05][YHa97].

## 4.2 The Ignition System

As mentioned in Chapter 2, turning on a natural gas MT is a complicated operation that can take up to 10 minutes (especially for older models). In the case of the Elliott MT, before supplying electrical energy to the grid, the whole system must be fed from an external source (a battery, for example) in order to accelerate the turbine to a certain speed that permits electrical energy generation.

An important aspect related to ignition is that the air flow to the combustor increases with the turbine speed [AHm04][RWK83][TAo83]. Ignition will always occur when air and fuel are sufficient to generate an appropriate air-to-fuel ratio [AHm04][EGS69][RFS00]. That is why fuel flow is imposed as a function of turbine speed.

As explained in Chapter 2, in order to achieve the optimum fuel-to-air ratio, certain parameters such as environmental temperature or atmospheric pressure must be previously known. Any error in measuring the turbine speed could result in a compromised fuel-to-air ratio [AHm04][RWK83]. In other words, ignition will not be carried out until the correct air-to-fuel ratio has been obtained. In most cases this is done based on experience [AHm04][EGS69] [RFS00].

In the case of the Elliott turbogenerator, after ignition has occurred, acceleration of the machine is rapidly increased and thus the air flow will increase. In this situation, the exhaust gas temperature will rise proportionally with air flow and turbine speed. That is why a controlled acceleration rate will be capable of providing cooler exhaust gases which will finally result in a smoother starting-up sequence and energy savings [AHm04][TAo83].

Coming back to the ignition process, after environmental temperature and atmospheric pressure have been determined, the gas turbine will be accelerated to a preset value in such a way that enough air arrives to the combustor [AHm04][EGS69][RFS00]. This acceleration rate of the turbines is fixed based on the fuel heating value and the above-mentioned environmental conditions. Acceleration is obtained through an external starter electrical motor (see Figure 4.1) which can be fed either from a battery or directly from the main network.

During the next stage of the process, the ignitor is activated, and then a constant fuel flow is provided to the combustor. This is maintained until the correct air-to-fuel ratio is attained and ignition occurs [AHm04][RFS00] [TAo83].

In the next stage, fuel flow is increased non-decreasingly [AHm04][EGS69] until the moment at which the MT is capable of driving a load. As in the case of the Capstone system, this one determines whether the lighting-off has taken place or not based on the gas exhaust temperature sensor reading [AHm04][EGS69][RWK83].

After the turbine has begun to drive the load (the electrical generator), its rotational speed is controlled and the outer electrical feeding is removed. The next stage consists of the fuel ramping up until the turbine is capable of actuating the electrical generator by itself. The MT is accelerated up to a speed of approximately 96,000–98,000 rpm. If the environmental temperature is lower, the turbine speed will be usually greater and thus its efficiency will be better.

The combustor can have one or more fuel orifices. In the second case, these orifices must communicate with the fuel source or sources [AHm04][RFS00][TAo83].

When the operation of the turbogenerator is idled, the fuel is delivered to the combustor at a constant flow rate [AHm04][EGS69].

When ignition does not occur (at a constant fuel flow rate) after a certain time period, the fuel flow is ramped up until ignition occurs. If this does not occur, the whole system is reset and purged [AHm04][RFS00]. For a better understanding of the interdependency between the air flow, fuel flow, and maximal speed achieved through an external grid connection, please refer to Figure 4.2 [AHm04][EGS69].

In Figure 4.2, I, II, III, IV, and V represent different time moments. In I, the ignitor is activated, in II the fuel is delivered to the combustor at a constant rate and after a certain period is ramped up until the turbine begins to drive the electrical generator [AHm04][RWK83][TAo83]. After ignition takes place, the air flow increases abruptly, as does the fuel flow, in order to accelerate the turbine to a self-sustainable speed [AHm04][EGS69]. For a better understanding of the ignition process, please refer to Figure 4.3 [AHm04][EGS69] [RWK83].

For more information on the Elliott ignition method, please refer to [AHm04][EGS69][RWK83][TAo83].

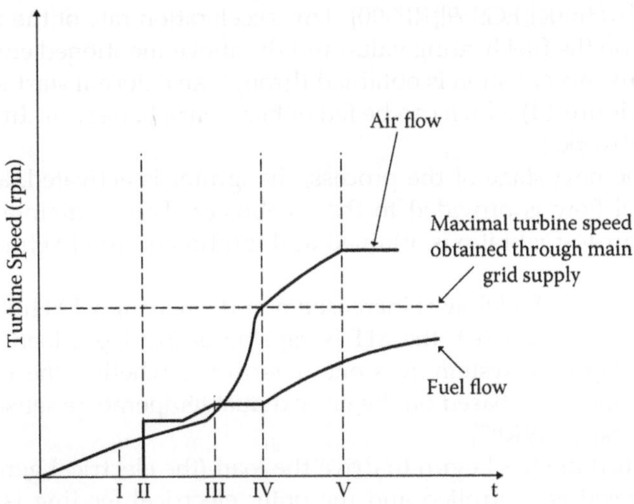


**FIGURE 4.2**
The approximate interdependency between the fuel flow, air flow, and maximal turbine speed achieved through the main grid supply. (From A. Hartzheim, Method for ignition and start up of a turbogenerator. U.S. Patent 6,766,647, issued July 27, 2004, available online: http://www.uspto.gov [accessed March 19, 2013]. With permission.)

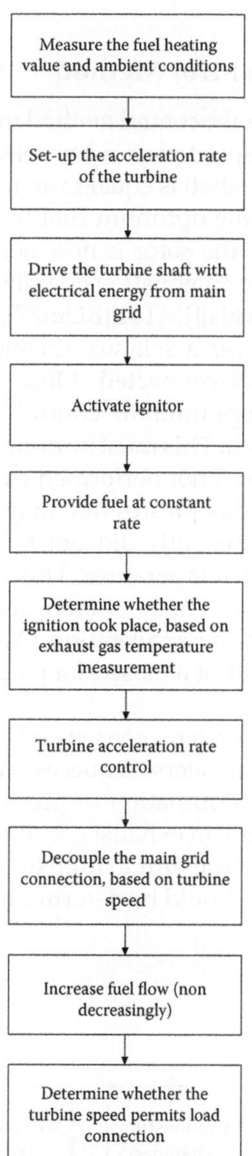


**FIGURE 4.3**
Approximate flowchart of the ignition process for an Elliott natural gas MT. (From A. Hartzheim, Method for ignition and start up of a turbogenerator. U.S. Patent 6,766,647, issued July 27, 2004, available online: http://www.uspto.gov [accessed March 19, 2013]; E.G. Smith et al., Automatic starting and protection system for a gas turbine. U.S. Patent 3,470,691, issued October 7, 1969, available online: http://www.uspto.gov [accessed March 20, 2013]; and R.W. Kiscaden et al., System and method for accelerating and sequencing industrial gas turbine apparatus and gas turbine electric power plants preferably with a digital computer control system. U.S. Patent 4,380,146, issued April 19, 1983, available online: http://www.uspto.gov [accessed March 20, 2013]. With permission.)

## 4.3 The Acceleration Control Method

After ignition has taken place, the control method for acceleration of the turboalternator to the synchronous speed plays a key role. Synchronous speed means that the rotation period of the shaft is equal to an integral number of AC cycles.

At the moment in which the optimum fuel-to-air ratio has been attained and the ignition took place, the rotor is now actuated by a combination of the power resulting from the combustion chamber and the power generated by the external battery [AAH04][JMT01][RUm77]. This combination is maintained until the shaft achieves a self-sustainable speed. At that point, the external power source is disconnected. Once the self-sustainable speed has been reached, it is important to control the acceleration until the synchronous speed is attained. This must be done in parallel with the exhaust gas temperature control, and if not performed could result in damage to the entire system due to high temperatures that may result.

Previous MT applications usually did not have a closed-loop control of acceleration and exhaust gas temperature. This could have provoked either overheating of the combustion chamber (and possibly damage to the entire MT) or incorrect control of the acceleration [AAH04][JMT01][KRC95]. The Elliott MT makes use instead of a closed-loop control method, as shown in Figure 4.4 [AAH04][JMT01].

The acceleration control system consists of a compressor, an annular combustor, and two PID controllers. A successful control method of such a turboalternator requires determination of acceleration rates (collected and presented in table format) and an exhaust gas temperature as low as possible in order to avoid exposure of the unit to high temperatures [AAH04][JMT01][RUm77]. Acceleration rates should be determined based on three criteria:

- Inlet temperature
- Engine speed
- Exhaust gas temperature

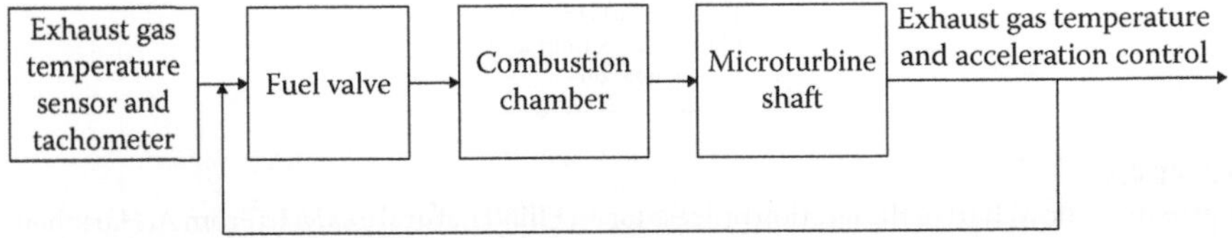


**FIGURE 4.4**
The Elliott acceleration control system. (From A. Hartzheim, Multiple control loop acceleration of turboalternator after reaching self-sustaining speed previous to reaching synchronous speed. U.S. Patent 6,834,226, issued December 21, 2004, available online: http://www.uspto.gov [accessed March 20, 2013]; and J.M. Teets et al., Electricity generating system having an annular combustor. U.S. Patent 6,314,717, issued November 13, 2001, available online: http://www.uspto.gov [accessed March 21, 2013]. With permission.)

After the external source has been cut off, the following operations must be carried out [AAH04][JMT01]:

- Switch on the timing device
- Monitor the rotor speed, exhaust gas temperature, inlet temperature, and acceleration rate
- Request the suitable acceleration rate from the created table
- Activate the PID controller to request a proper fuel valve position based on the exhaust gas temperature
- Select the valve position based on the least amount of gas entering the combustion chamber
- Exit the control loop if synchronous speed is achieved
- Exit the control loop if synchronous speed is not achieved within a certain time limit and thus the combustion chamber has to be purged and the whole system restarted
- Repeat all the previous steps in the same order until synchronous speed is achieved

For an illustration of the control system, please refer to Figure 4.5 [AAH04][JMT01]. This figure presents the relation between the various parts of the Elliott acceleration system, and Figure 4.5 presents a flowchart of the software that has been developed to manage this system.

As can be observed in Figure 4.4, the control loop uses information on the exhaust gas temperature and acceleration (which is taken from special dedicated sensors) and further transmits a signal (along with the information from the tachometer and another exhaust gas temperature sensor) to choose the optimum amount of natural gas to be directed to the combustion chamber. The resulting thermal energy will drive the MT shaft and finally the electrical generator. Note that this control system is turned on after the turbogenerator has reached a self-sustainable speed and the outer electrical source has been disconnected. The basic principle according to which this control system is developed can be described through Figure 4.5, as follows: if the tachometer has "sensed" that the rotor has reached the synchronous speed the system will be exited. If this did not happen, an acceleration rate is requested based on the above-mentioned table [AAH04][JMT01]. According to [AAH04][RUm77], typical values for the acceleration rate could be between 10,000 rpm and 70,000 rpm if the inlet temperature is in the range between –29°C and 60°C, and thus the exhaust gas temperature would be in the interval 149°C–649°C.

During the next stage, the control system will use one of the PIDs to impose an optimum position of the fuel valve based on the requested acceleration rate, and the other PID will do the same based on a pre-stored moderate exhaust gas temperature and on the actual exhaust gas temperature [AAH04][JMT01]. These two values should correspond. The system will also choose

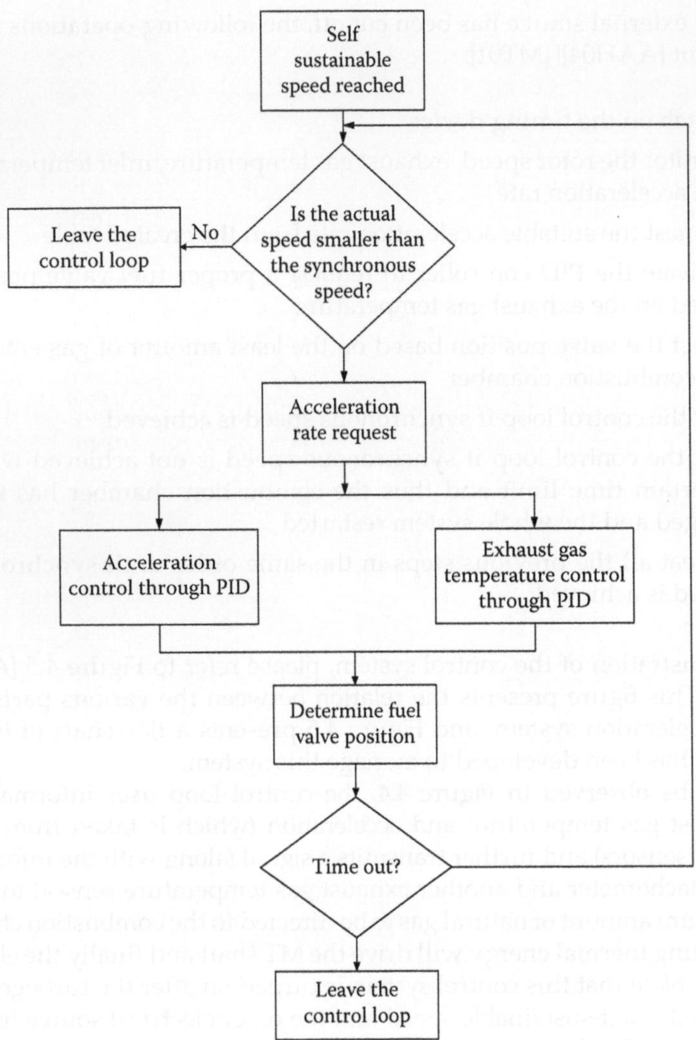


**FIGURE 4.5**
General flowchart of the control software. (From A. Hartzheim, Multiple control loop acceleration of turboalternator after reaching self-sustaining speed previous to reaching synchronous speed. U.S. Patent 6,834,226, issued December 21, 2004, available online: http://www.uspto.gov [accessed March 20, 2013]; J.M. Teets et al., Electricity generating system having an annular combustor. U.S. Patent 6,314,717, issued November 13, 2001, available online: http://www.uspto.gov [accessed March 21, 2013]; and K.R. Carr et al., Engine starting system utilizing multiple controlled acceleration rates. U.S. Patent 5,430,362, issued July 4, 1995, available online: http://www.uspto.gov [accessed March 21, 2013]. With permission.)

that position of the fuel valve which assures a lesser amount of natural gas in the combustion chamber [AAH04][KRC95][RUm77]. Using this type of control, the exhaust gas temperature should not attain a value that would damage the unit.

After the position of the fuel valve has been precisely determined, a special dedicated chronometer will be utilized to verify if the control loop has timed out. This comparative value is chosen based on real experiments and based on the time period needed by the MT to be turned on [AAH04][JMT01]. When it takes too long to reach synchronous speed, the system control is exited and the turbogenerator is shut down and purged [AAH04][RUm77]. If this is not the case, the whole process is repeated. Figure 4.6 shows the approximate representation of the interdependence between rotor speed and time. In this figure, 0 identifies the moment at which the self-sustainable speed has been achieved and *t*1 identifies the moment at which the synchronous speed has been reached.

Figure 4.6 represents proof of the fact that when using this type of control, no important or unexpected variations will appear in the evolution of the shaft speed which could lead to damage of the unit.

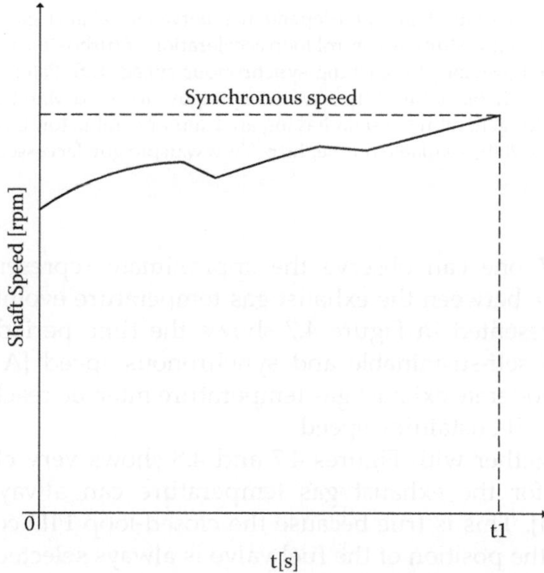


**FIGURE 4.6**
Approximate representation of the interdependence between rotor speed and time. (From A. Hartzheim, Multiple control loop acceleration of turboalternator after reaching self-sustaining speed previous to reaching synchronous speed. U.S. Patent 6,834,226, issued December 21, 2004, available online: http://www.uspto.gov [accessed March 20, 2013]; and J.M. Teets et al., Electricity generating system having an annular combustor. U.S. Patent 6,314,717, issued November 13, 2001, available online: http://www.uspto.gov [accessed March 21, 2013]. With permission.)

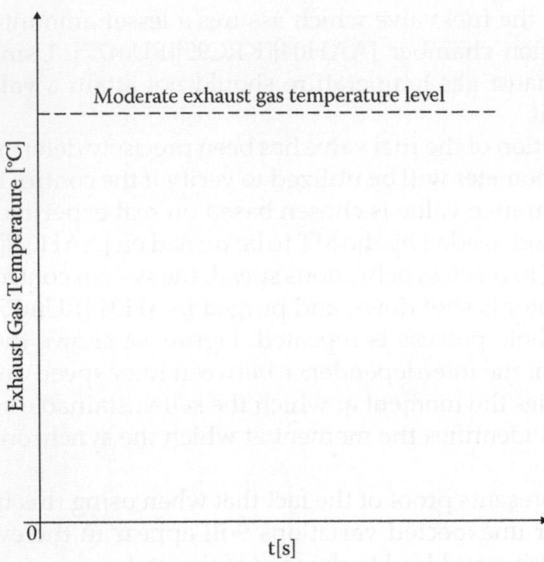


**FIGURE 4.7**
Approximate representation of the interdependence between exhaust gas temperature and time. (From A. Hartzheim, Multiple control loop acceleration of turboalternator after reaching self-sustaining speed previous to reaching synchronous speed. U.S. Patent 6,834,226, issued December 21, 2004, available online: http://www.uspto.gov [accessed March 20, 2013]; and J.M. Teets et al., Electricity generating system having an annular combustor. U.S. Patent 6,314,717, issued November 13, 2001, available online: http://www.uspto.gov [accessed March 21, 2013]. With permission.)

In Figure 4.7, one can observe the approximate representation of the interdependence between the exhaust gas temperature evolution and time. The relation presented in Figure 4.7 shows the time period between the achievement of self-sustainable and synchronous speed [AAH04][JMT01][KRC95]. The moderate exhaust gas temperature must be reached before the MT attains the self-sustaining speed.

Figure 4.6 together with Figures 4.7 and 4.8 shows very clearly that the desired value for the exhaust gas temperature can always be selected [AAH04][JMT01]. This is true because the closed-loop PID control systems are stable, and the position of the fuel valve is always selected based on the situation in which a smaller amount of natural gas is diverted in the combustion chamber. For instance, if one of the PIDs requests a higher acceleration rate, then the other PID whose aim is to obtain a smaller value for the gas exhaust temperature will automatically solicit less fuel.

Figure 4.9 presents a situation in which the moderate exhaust gas temperature level is never reached. Here, 0 identifies the moment at which the self-sustaining speed has been attained and *t*1 identifies the moment at which the synchronous speed is achieved.

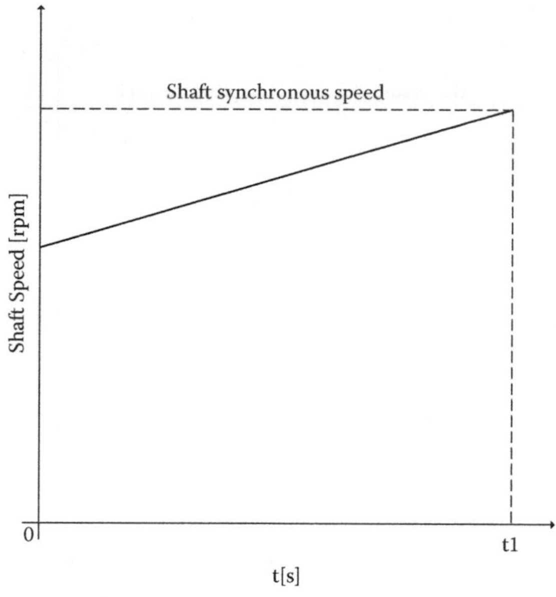


**FIGURE 4.8**
Approximate representation of the interdependence between shaft synchronous speed and time. (From A. Hartzheim, Multiple control loop acceleration of turboalternator after reaching self-sustaining speed previous to reaching synchronous speed. U.S. Patent 6,834,226, issued December 21, 2004, available online: http://www.uspto.gov [accessed March 20, 2013]; and J.M. Teets et al., Electricity generating system having an annular combustor. U.S. Patent 6,314,717, issued November 13, 2001, available online: http://www.uspto.gov [accessed March 21, 2013]. With permission.)

Figures 4.8 and 4.9 demonstrate that the system will always select the appropriate acceleration rate due to the fact that when the exhaust gas temperature is at a maximum, the PID command coordinating this temperature will be overridden by the PID command coordinating the acceleration rate, which in this situation assures a smaller amount of fuel in the combustion chamber [AAH04][JMT01].

Figures 4.7, 4.8, and 4.9 present only an extreme scenario when both PIDs are soliciting fuel valve position (and implicitly an acceleration rate) simultaneously, which must assure the smallest quantity of natural gas in the combustion chamber. In most cases, the fuel valve position is not imposed only by a single PID during the time interval between achieving the self-sustainable speed and the synchronous speed, but by both [AAH04][JMT01]. This mode of operation avoids practically any situation in which abrupt variations of the rotor speed or overheating of the combustion chamber can arise [AAH04][JMT01].

For further details on this subject, please refer to [AAH04][JMT01][KRC95][RUm77].

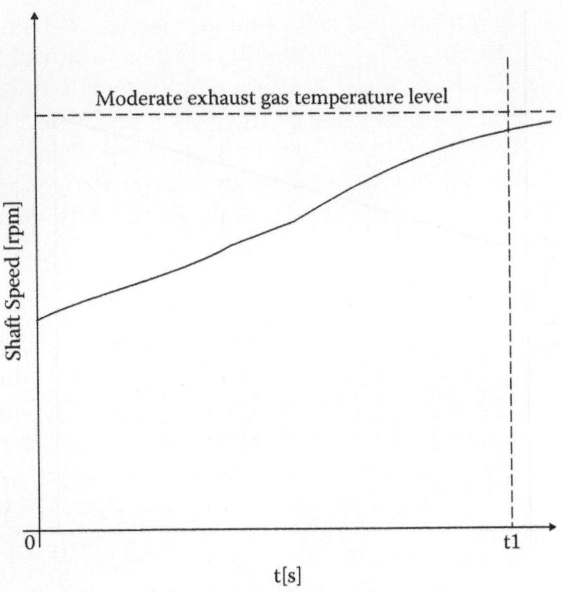


**FIGURE 4.9**
Approximate representation of the interdependence between moderate exhaust gas temperature and time. (From A. Hartzheim, Multiple control loop acceleration of turboalternator after reaching self-sustaining speed previous to reaching synchronous speed. U.S. Patent 6,834,226, issued December 21, 2004, available online: http://www.uspto.gov [accessed March 20, 2013]; J.M. Teets et al., Electricity generating system having an annular combustor. U.S. Patent 6,314,717, issued November 13, 2001, available online: http://www.uspto.gov [accessed March 21, 2013]; and R. Uram, Accurate, stable and highly responsive gas turbine startup speed control with fixed time acceleration especially useful in combined cycle electric power plants. U.S. Patent 4,010,605, issued March 8, 1977, available online: http://www.uspto.gov [accessed March 20, 2013]. With permission.)

## 4.4 The Recuperator Structure

As in the case of the Capstone MT, the Elliott recuperator has a heat exchange structure where the compressor discharge air is placed in indirect contact with the turbogenerator exhaust gases before being dispersed in the environment [DWD04][SFo80][SFr80].

In the case of the first recuperator version developed by Elliott, there was the problem of insufficient heat exchange between these two mediums [DWD04][SFr80]. This means that there was a problem of not efficiently utilizing the MT exhaust gases before being placed into indirect contact.

The solution adopted was to encapsulate the recuperator [DWD04][SFo80]. In this way, a lesser amount of heat is lost before the heat recovery which represents the principal goal of the recuperator [DWD04][FHV59].

Another advantage of this solution is that it provides an evenly distributed back pressure, which is synonymous with a more uniform flow distribution and improved efficiency of operation [DWD04][FHV59]. Using this type of approach, an optimized discharge gas passage of the recuperator is obtained, and hence the heat transfer between fluids is greatly improved. Such a solution permits efficient use of the MT for cogeneration. In this situation, the recuperator should have a bypass valve that not only could power the cogeneration system but also could adjust MT efficiency in general [DWD04][SFo80].

Coming back to the recuperator structure, the cylindrically shaped enclosure of the recuperator will force the exhaust gases after they have completed indirect contact to surround the recuperator and thus to build an insulation coating, recapturing part of the heat that has been initially lost through the exhausting process [DWD04][FHV59]. In this way, the possibility of increasing the temperature of the other MT components arises, this being one of the main factors contributing to improving the efficiency of the turbogenerator operation [DWD04][SFo80][SFr80]. The direct consequence of this insulation is that the temperature difference on the outer diameter of the recuperator will be only 38°C, a greater difference contributing to, among other things, metal fatigue. For an illustration of the Elliott recuperator structure, please refer to Figure 4.10.

As can be seen in Figure 4.10, the MT exhaust gases flow in the recuperator initially from left to right and then through the discharge passage defined

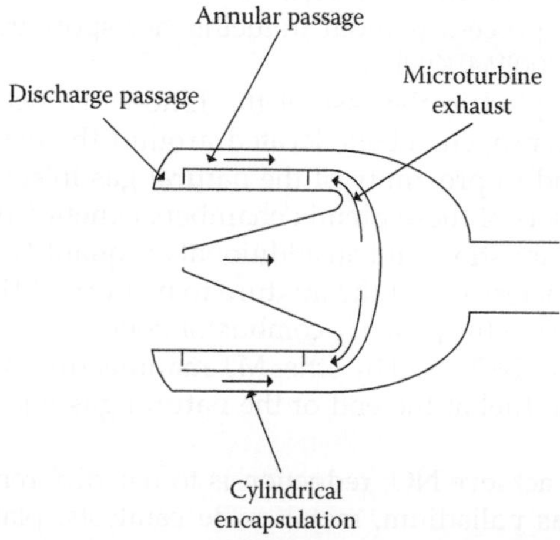


**FIGURE 4.10**
Approximate view of the Elliott recuperator structure. (From D.W. Dewis, Recuperator configuration. U.S. Patent 6,832,470, issued December 21, 2004, available online: http://www.uspto.gov [accessed March 21, 2013]; and S. Forster et al., Vehicular gas turbine installation with ceramic recuperative heat exchanger elements arranged in rings around compressor, gas turbine and combustion chamber. U.S. Patent 4,180,973, issued January 1, 1980, available online: http://www.uspto.gov [accessed March 22, 2013]. With permission.)

by the cylindrical encapsulation, and finally through the annular passage before being dispersed in the environment. The exhaust gases have a temperature of approximately 260°C, and that is why in general the MT applications are suitable for cogeneration. In this case one could use these gases to heat the water from a boiler.

For further details on the Elliott recuperator structure, please refer to [DWD04][FHV59][SFo80][SFr80].

## 4.5 The $NO_x$ Reduction System

The Elliott $NO_x$ reduction system basically minimizes the NO and the $NO_2$ [JBI04][JLi00][RPL81].

One of the most important problems in minimizing the pollutant emissions of a natural gas MT is related to the pollutants resulting from the fuel burning at low flame (like the CO) and those resulting from the fuel burning at high flame (like the $NO_x$). That is why the simultaneous optimization of these two pollutants must always be associated with two conflicting objective functions, as described in Chapter 3.

Another important aspect is represented implicitly by the design of the combustor, which plays a key role, especially with high flame temperatures. This is because the mixture between air and fuel has to be even during the entire combustion process without inducing hot spots in the combustion chamber [JBI04][JLi00][MJK92].

The solution adopted in the case of the Elliott MT was to create a certain number of premix chambers, located around the external housing of the combustors and in proximity of the natural gas injectors [JBI04][JLi00]. In this way, the inlets of these premix chambers benefit from a rich air–fuel mixture before combustion, and an additional air quantity is injected which improves the characteristics of the mixture to be burned [JBI04][JLi00]. This burning takes place in the primary combustor zone.

Other methods for $NO_x$ used by other MT manufacturers consisted of mixing the air and the fuel at the end of the natural gas injector [JBI04][JLi00][RPL81].

Another way to achieve $NO_x$ reduction is to use different catalysts in the combustor, such as palladium, metal oxide catalysts, platinum, cobalt, or nickel [JLi00].

The standard configuration of the Elliott combustor includes a dam that has the role of reducing $NO_x$ emissions resulting from the combustion chamber. For a diagram of this combustor, please refer to Figure 4.11.

The fuel and air intakes in Figure 4.11 have the role of assuring a recirculation zone for the continuous fuel combustion. The combustion products are

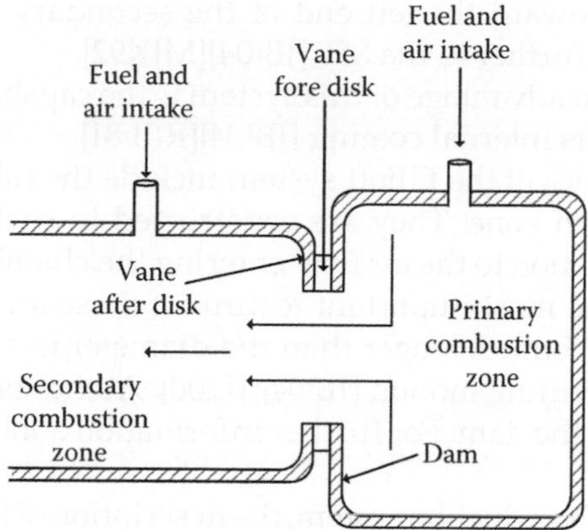


**FIGURE 4.11**
Approximate diagram of the Elliott combustor. (From J.B. Ingram, MT with auxiliary air tubes for $NO_x$ emission reduction. U.S. Patent 6,729,141, issued May 4, 2004, available online: http://www.uspto.gov [accessed March 23, 2013]; and J. Lipinski et al., Low $NO_x$ conditioner system for a MT power generating system. U.S. Patent 6,125,625, issued October 3, 2000, available online: http://www.uspto.gov [accessed March 23, 2013]. With permission.)

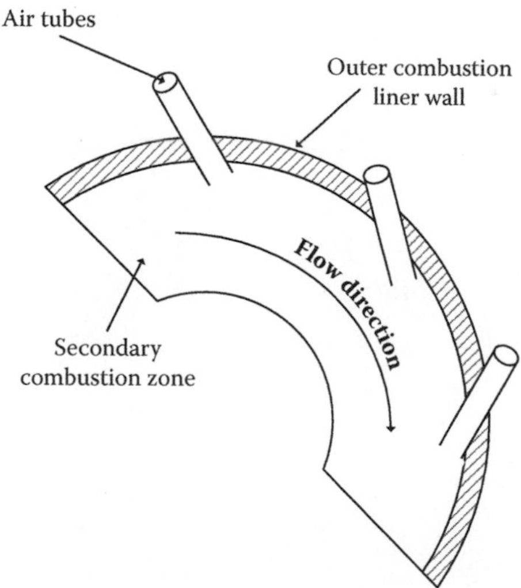


**FIGURE 4.12**
A fragmentary sectional view of the Elliott combustor. (From J.B. Ingram, MT with auxiliary air tubes for $NO_x$ emission reduction. U.S. Patent 6,729,141, issued May 4, 2004, available online: http://www.uspto.gov [accessed March 23, 2013]; and J. Lipinski et al., Low $NO_x$ conditioner system for a MT power generating system. U.S. Patent 6,125,625, issued October 3, 2000, available online: http://www.uspto.gov [accessed March 23, 2013]. With permission.)

normally directed toward the left end of the secondary combustion zone, which diverts them further to the MT [JBI04][MJK92].

Another important advantage of this system is the capability of prolonging combustor life and its internal coating [JBI04][RPL81].

Other characteristics of the Elliott system include the tubes that divert the air to the combustion zone, They are constructed in such a way that they induce a swirling motion to the air flow entering the chamber [JBI04][MJK92] (see Figure 4.12). The most important feature of these air tubes is that they have to be at least 1.5 times longer than the diameter in order to induce the above-mentioned swirling motion [JBI04][JLi00]. At the same time they have to be mounted after the dam. For further information about this, please refer to [JBI04][JLi00].

With the Elliott $NO_x$ reduction system, the description of the most important components of the TA 100 MT comes to an end. Its electrical performances will be thoroughly investigated in Chapter 6.

# 5

# *Power Converter Circuits Used for Grid Connection*

Theoretical possibilities for grid connection of a natural gas microturbine (MT) are explored in detail in [ORNL-- ] and [Ko04]. The modeling, simulation, and connection to the grid of a MT are investigated as well in [AC05][APA07][AW04][FDA04][GDL06][GP06][GuDL06][NL03][RNM06][SCC08][YT09][YWS03].

Many of the possibilities mentioned in these works are not used in real life (i.e., the matrix convertor). The cycloconverter is normally used for generators with a high number of pole pairs. In an overwhelming number of cases, classical AC/AC convertors are employed. Sections 5.1 and 5.2 explain this further.

## 5.1 Power Converter Circuits Used for C30 and C60

The core of this grid connection system consists of a high-frequency inverter synchronously connected to the permanent magnet generator of the gas MT, a low-frequency load inverter and a central processing unit which controls the voltage, current, and frequency of both inverters (see Figure 5.1).

Figure 5.2 is a diagram of the power control system. The approximate functional block diagram of the power control system is shown in Figures 5.3a and 5.3b [BE01][BEL02].

Coming back to Figure 5.1, the controller **8** consists of two bidirectional converters [BE01][BEL02], a low-frequency load inverter **5**, and the generator inverter **4**. The controller **8** receives electrical power **7** from the grid through the AC filter **6** or from a battery through battery control electronics **9**. The generator inverter **4** starts the turbine through the power head **2** (using the permanent magnet generator as motor) with electricity either from the grid or battery, and then the inverter **5** begins to produce AC power using the output power from the generator inverter **4** to bring power from the high-speed permanent magnet turbogenerator [BE01][BEL02]. The controller **8** regulates the fuel flow to the combustor through the fuel control valve **10** [BE01][BEL02].

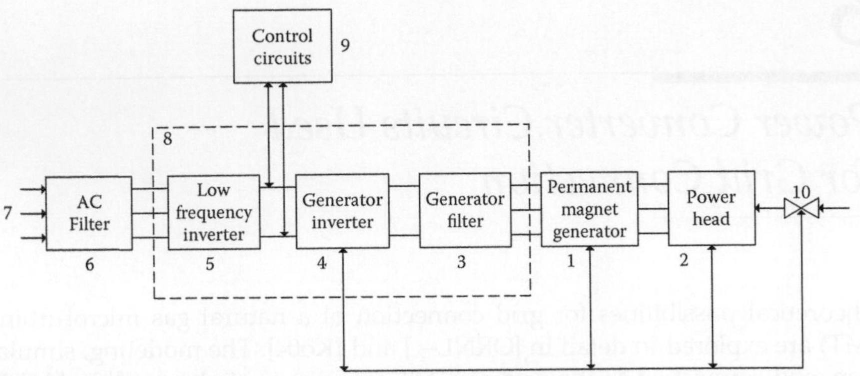


**FIGURE 5.1**
The interface between the MT controller and the permanent magnet generator. (From R. Bosley et al., Turbogenerator power control system. U.S. Patent 20010052704, issued December 20, 2001, available online: http://www.uspto.gov [accessed March 20, 2013]; and R.W. Bosley et al., Turbogenerator power control system. U.S. Patent 6,495,929, issued December 17, 2002, available online: http://www.uspto.gov [accessed March 20, 2013]. With permission.)

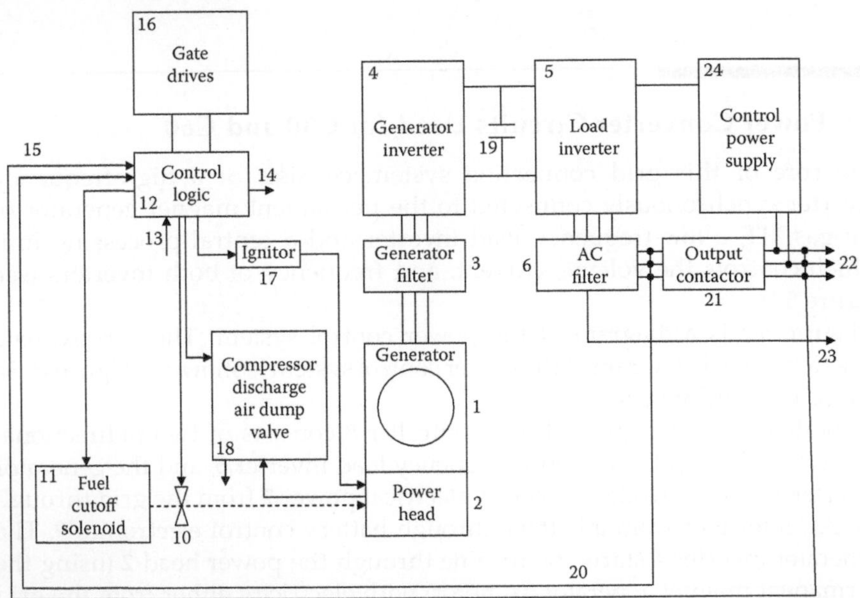


**FIGURE 5.2**
Connections for the power control system. (From R. Bosley et al., Turbogenerator power control system. U.S. Patent 20010052704, issued December 20, 2001, available online: http://www.uspto.gov [accessed March 20, 2013]; and R.W. Bosley et al., Turbogenerator power control system. U.S. Patent 6,495,929, issued December 17, 2002, available online: http://www.uspto.gov [accessed March 20, 2013]. With permission.)

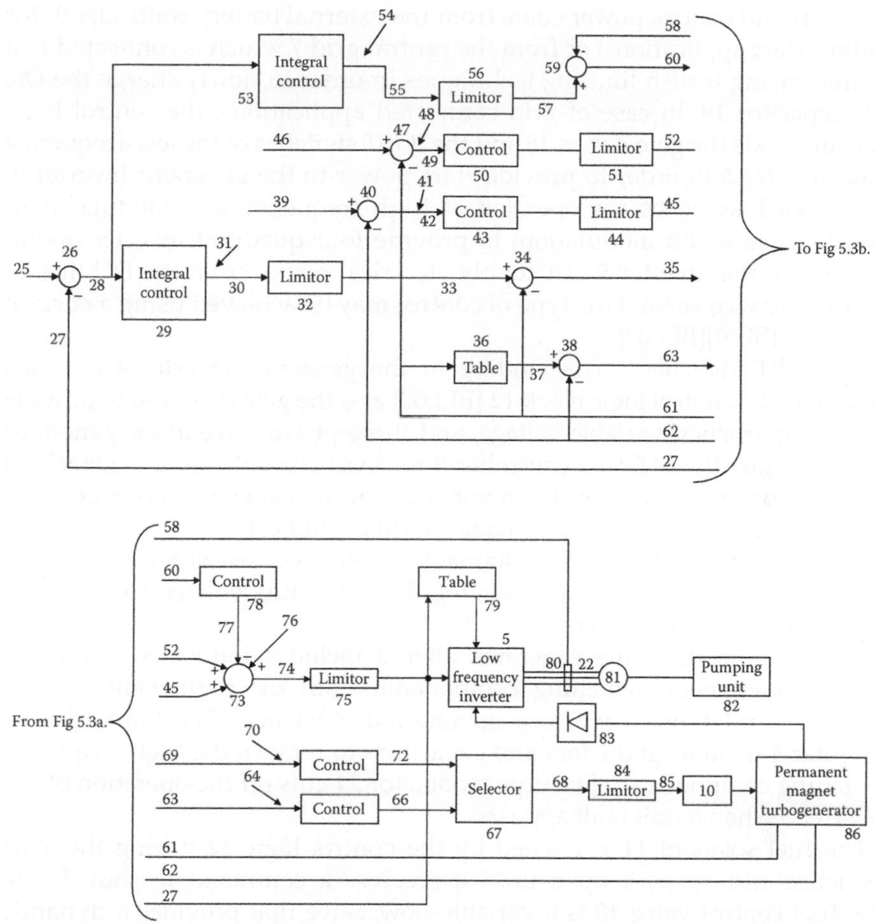


**FIGURE 5.3**
(a,b) Approximate functional block diagram of the power control system. (Parts a and b: From R. Bosley et al., Turbogenerator power control system. U.S. Patent 20010052704, issued December 20, 2001, available online: http://www.uspto.gov [accessed March 20, 2013]; and R.W. Bosley et al., Turbogenerator power control system. U.S. Patent 6,495,929, issued December 17, 2002, available online: http://www.uspto.gov [accessed March 20, 2013]. With permission.)

The controller **8** can be seen in more detail in Figure 5.2. It is made up of insulated gate bipolar transistor (IGBT) gate drives **16**, a control logic block **12**, a generator inverter **4**, a permanent magnet generator filter **3**, a DC bus capacitor **19**, a low-frequency inverter **5**, an AC filter **6**, an output contactor **21**, and a control power supply **24**. Through the control logic block **12**, power is also provided to the fuel cutoff solenoid **11**, to the control fuel valve **10**, and also to the ignitor **17**. The battery controller **9** is connected directly to the DC bus. The control logic **12** receives a temperature signal **13**, a voltage signal **15**, and current signal **20** while it provides a relay drive signal **14**.

The start and control power come from the external battery controller **9** (for battery start applications) or from the central grid **7** which is connected to a rectifier using inrush limiting techniques in order to slowly charge the DC bus capacitor **19**. In case of grid-connected applications, the control logic **12** commands the gate drives **16** and the IGBT switches of the low-frequency load inverter **5** in order to provide start power to the generator inverter **4**. These IGBT switches are operated at high frequency and modulated in PWM (pulse width modulation) to provide four-quadrant inverter operation where the inverter **5** can supply electrical power from the DC link to the grid, or vice versa. This type of control may be achieved using a current regulator [BE01][BEL02].

The IGBT switches corresponding to the generator inverter **4** are also driven by the control logic block **12** [BEL02] and the gate drives **16** to provide variable frequency, variable voltage, and three-phase drive to the generator motor to start the MT. The controller **8** receives current signal feedback **20** through current sensors in the moment when the generator has been commanded to finalize the starting sequence [BE01][BEL02].

When the MT reaches self-sustainable speed, the inverter **4** changes its operation mode in order to provide regulated DC link voltage and to boost the generator output voltage.

At the same time, the generator filter **3** includes inductors to remove the high-frequency switching components from the permanent magnet generator and thus to increase operational efficiency [BE01][BEL02]. The AC filter **6** includes inductors and capacitors to remove the high-frequency switching components. The output contactor **21** cuts off the operation of the inverter **5** when a unit fault appears.

The fuel solenoid **11** is opened by the control logic **12** during the start sequence and keeps it open until it receives a command to shut down. The fuel control valve **10** is a variable flow valve that provides a dynamic regulating range, allowing a minimum fuel quantity during start and a maximum fuel quantity at full load. The ignitor **17** is similar to a spark plug and, normally, is operated only during the starting sequence [BE01][BEL02].

In the case of a stand-alone operation of the MT, this is started through an external DC converter that will boost the voltage from the external unit and connects directly to the DC link. The low-frequency load inverter **5** can be used now as a constant frequency and as a constant voltage source. The most important advantage is that this system facilitates at output a variable frequency and voltage.

The block diagram of the entire system is presented in Figures 5.3a and 5.3b [BE01][BEL02]. The three primary control loops used to regulate the gas MT engine are visible in Figure 5.3b. These control loops are the power control loop **76**, the turbogenerator speed control loop **70**, and the exhaust gas temperature control loop **64**. The speed control loop **70** commands the fuel output to the MT fuel control **10** in order to regulate the rotating speed of the shaft [BE01][BEL02].

The exhaust gas temperature control loop **64** commands the fuel output to the MT fuel control **10** in order to regulate the operating temperature of the turbogenerator. The minimum fuel command **68** is selected by the selector **67** which chooses the smallest signal from the speed control loop **70** and the MT exhaust gas temperature control loop **64** [BE01][BEL02].

There is a possibility of turbogenerator overspeeding when a stored thermal energy device is used (i.e., a recuperator). In this case, the maximum speed control loop **48** (Figure 5.3a) varies the frequency to the low-frequency generator **5**. The frequency offset signal **52** (Figure 5.3b) is produced by the limitor **51** (Figure 5.3a). If the load is an engine, its speed can be varied to control maximum or transient exhaust gas temperature by a maximum transient exhaust gas temperature control loop **41**. The frequency offset signal **45** is produced by the limitor **44** [BE01][BEL02]. An important advantage of this is that the turbogenerator and its power control system can be moved from a group of one or more loads to another group of loads without it being necessary to adjust any of the control system parameters.

The power control system is capable of automatically adapting itself to powering any number of loads from one to the maximum number permitted by the power level available from the MT, and it can tolerate all of the power peaks of the loads either at the same time or out of phase [BE01][BEL02].

As shown in Figures 5.3a and 5.3b, the average frequency **25** necessary to produce three-phase electrical power generated by the low-frequency inverter **5** is compared in the summer **26** with the instantaneous frequency **27** resulting from the inverter **5**. The difference between these frequency values will generate the error signal **28** that will be utilized as the input for the MT speed command control loop **31** and power command control loop **54**. When the average over time of the error signal **28** is zero, the power required by the loads is equal to the power generated by the MT [BE01][BEL02].

The speed command control loop **31** and the proportional integral control **29** generate a recommended speed signal **30** for the MT, and this should produce a level of electrical power equal to that required by the load. This recommended speed signal **30** is limited by the limitor **32** to a maximum value corresponding to the maximum safe operating speed of the turbogenerator and is also limited by the same limitor to a minimum value corresponding to the minimum speed at which the MT can operate without power output [BE01][BEL02].

The proportional integral control **53** of the power command control loop **54** imposes a recommended power consumption level signal **55** for the loads that should be equal to the power level generated by the unit. This recommended power consumption level signal **55** is limited by the limitor **56** to zero when the load circuit breakers are open [BE01][BEL02].

The output signal **33** belonging to the speed command control loop **31** represents a speed command **33** to the MT. This speed command **33** is compared in the summer **34** to the real turbogenerator speed feedback signal **61** coming from the unit. The error signal **35** between these two speed values is supplied to the proportional integral control **71** of the speed control loop **70** in order to generate an optimal fuel flow signal **72** [BE01][BEL02].

The look-up table **36** is used together with the real speed feedback signal **61** to impose a recommended exhaust gas temperature command **37** for the MT. This recommended exhaust gas temperature command **37** is compared in the summer **38** to the real turbine exhaust gas temperature feedback **62** to generate a computed exhaust gas temperature error signal **63**. This computed exhaust gas temperature error signal **63** is brought to the proportional integral control **65** in the turbine exhaust gas temperature loop **64** which will impose a recommended fuel flow signal **66** that normally should eliminate the temperature error [BE01][BEL02]. The selector **67** chooses the lowest signal from the exhaust gas temperature loop **64** and the speed control loop **70**, providing the lower signal to the limitor **84**, which limits the recommended fuel flow to a maximum value equal to that required to produce the maximum power that can be produced by the MT and to a minimum value below which the combustor will experience flame-out. The selected fuel flow value **85** is used afterward by the fuel control **10** to determine the required fuel flow rate to the combustor. The resulting speed feedback signal **61** and the exhaust gas temperature feedback signal **62** are measured in the turbogenerator and again utilized in the power control system [BE01][BEL02].

The output **57** resulting from the limitor **56** represents the low-frequency inverter **5** average power command that is then compared in the summer **59** to the real instantaneous power feedback signal **58** from the power sensor **80**. The resulting error signal **60** is afterwards utilized in the proportional integral control **78** to impose a recommended instantaneous inverter frequency signal **77** that normally eliminates the power error [BE01][BEL02].

The summer **47** compares the speed feedback signal **61** to the maximum safe speed feedback signal **46** to generate a speed error signal **49**. If the MT speed is greater than the maximum safe speed **46**, the proportional integral control calculates a recommended frequency increase signal **52** (which is then limited in the limitor **51**) in the low-frequency load inverter.

The exhaust gas temperature feedback signal **62** is compared to the maximum safe exhaust gas temperature signal **39** in the summer **40** to generate an error signal **42**. If the exhaust gas temperature of the unit is greater than the maximum safe temperature **39**, the proportional integral control **43** calculates a recommended frequency increase signal **45** (limited in the limitor **44**) in the low-frequency load inverter [BE01][BEL02].

Both inverter frequency reduction signals **52** and **45** are brought to the summer **73** which also receives the signal **77**. The resulting error signal **74** is limited **75** before going to the low-frequency load inverter **5**. This limited error signal controls the frequency of the inverter **5** and generates a frequency limit signal **27** to both summer **26** and the look table **79** which then calculates the inverter output voltage [BE01][BEL02].

Finally, the three-phase electrical power generated by the low-frequency load inverter **5** passes through the power sensor **80**. The signal **58** is utilized by the summer **59** to assure that the power from the inverter **5** corresponds to the turbogenerator power required to maintain the average frequency of the inverter **5**. The frequency corresponding to **5** can be set to 50 or 60 Hz.

Without the power control system described above or a similar one, the power requirements can vary over very short periods of time (a few seconds) by four times the average power required by the load. This results in an increased size of the MT which leads eventually to important centrifugal and thermal stresses, cyclical variations in operating speed, unstable operation, and low efficiency in operation [BE01][BEL02]. This type of power control system allows the MT to supply one or more varying loads without the need to adjust the fuel consumption, combustion temperature, or unit speed [BE01][BEL02].

For further information on the Capstone power conversion system, please refer to [BE01][BEL02].

## 5.2 Power Converter Circuits Used for TA-100

In TA-100 microturbines (MTs), the power control system is configured to conduct a synchronized operation not only when there is a single supply unit but also when a cluster of power supply units has to be coordinated [TFa09][WBH00]. This can occur either in stand-alone mode or connected to the main grid.

The present power conversion system refers to a cluster of units connected in parallel so that at least one turbogenerator is automatically driven at its rated parameters, and the others are operated in order to adjust the electrical power output [TFa09][WBH00].

The core of this power control system consists of an AC/DC converter for converting the voltage produced by the turbogenerator into DC voltage, a booster for boosting the converted DC voltage, and a DC/AC converter to transform the boosted DC current into AC voltage for the output. The load sharing adjuster of the inverter controls not only the booster but also the DC/AC converter to produce a constant output DC voltage from the booster, and at the same time to generate a constant AC voltage from the DC/AC converter when the current output from the DC/AC converter corresponds to or is less than a predetermined current which is less than the AC-rated current [TFa09][WBH00]. The load adjuster contributes also to the following [TFa09][WBH00]:

- The output of a DC voltage, which gradually decreases as the booster current increases
- The output of an AC voltage, which gradually decreases from the AC/DC converter when the current output from the DC/AC converter is more than a predetermined current

In this sense, the load adjusters of the MT cluster control the associated boosters to generate constant output DC voltages that differ from each other and control the corresponding DC/AC converters to produce constant output AC voltages which again differ from each other [TFa09][WBH00]. To achieve these objectives, each unit has to include [TFa09][WBH00]:

- A generator
- An inverter for converting the voltage produced by the generator into an AC voltage and for outputting this AC voltage
- An inverter control unit for controlling the inverter, consisting of a first synchronization controller operated in linkage mode with an external AC power supply in order to detect a voltage resulting from it, which will serve to control the inverter so that its output AC voltage is in phase with the voltage of the external AC power supply
- A connection box to connect the AC voltage output from the inverter to the bus
- A multiple power unit controller serving to coordinate the MT cluster to individually start and stop the units and, at the same time, to control the outputs of each unit in the cluster

A simplified block diagram illustrating the power conversion system of one unit can be seen in Figure 5.4 [TFa09][WBH00]. The simplified block diagram of the power conversion system for a cluster of TA-100 units can be seen in Figure 5.5 [TFA09][WBH00].

The scheme in Figure 5.4 includes, in addition to the power conversion system of a TA-100 unit, the MT itself **1** and a rotor generator **2** [TFa09][WBH00]. This is in fact a DC brushless generator having the stator around the rotor [TFa09][WBH00].

In the gas turbine generator, the gaseous fuel is supplied via a feeder (not shown here) to the combustion chamber, through a fuel control valve **20**. The combustion gas that passed through the MT blade is heat exchanged with the air compressed by the compressor blade at the regenerator and emitted outside. The compressed air, previously heated by the regenerator, is supplied to the combustion chamber and finally mixed with fuel to be burnt and thus to rotate the turbine at high speeds [TFa09][WBH00].

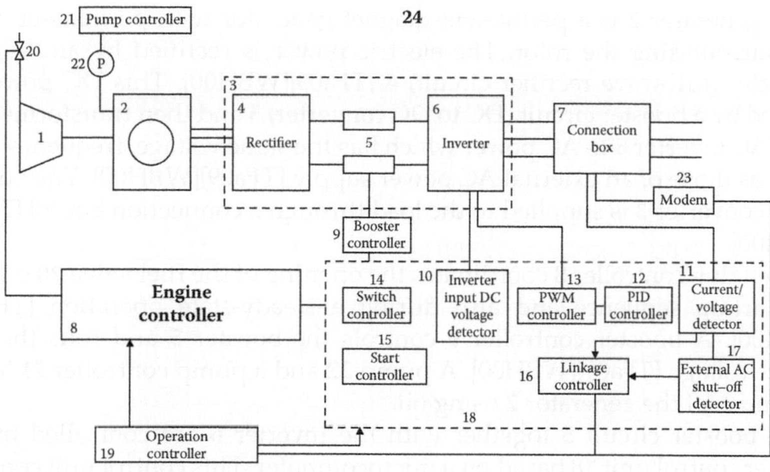


**FIGURE 5.4**
Approximate view of the simplified block diagram for the power conversion system of one TA-100 unit. (From T. Furuya et al., Electric power supply system. U.S. Patent 7,514,813, issued April 7, 2009, available online: http://www.uspto.gov [accessed March 22, 2013]; and W.B. Hall et al., Communications processor remote host and multiple unit control devices and methods for micropower generation systems. U.S. Patent 6,055,163, issued April 25, 2000, available online: http://www.uspto.gov [accessed March 22, 2013]. With permission.)

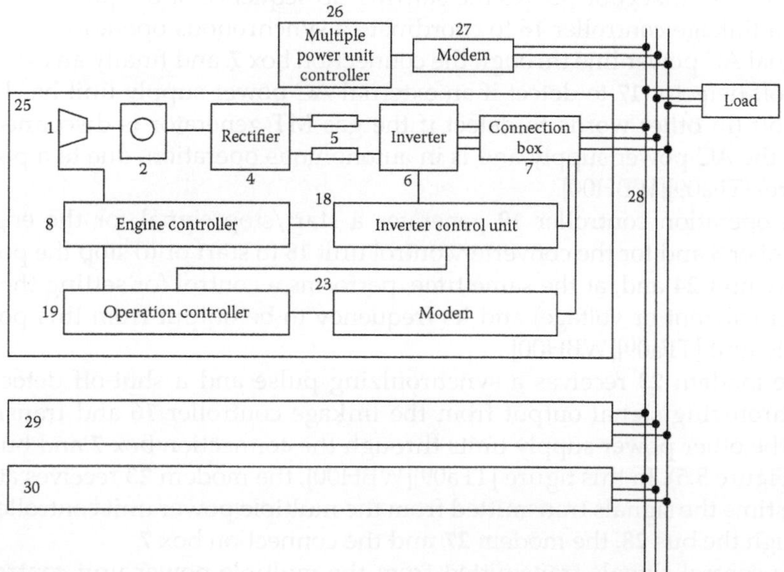


**FIGURE 5.5**
Approximate view of the simplified block diagram for the power conversion system for a cluster of Calnetix units. (From T. Furuya et al., Electric power supply system. U.S. Patent 7,514,813, issued April 7, 2009, available online: http://www.uspto.gov [accessed March 22, 2013]. With permission.)

The generator **2** is a permanent magnet generator with permanent magnets surrounding the rotor. The electric power is rectified by an AC/DC converter (full-wave rectifier circuit) **4** [TFa09][WBH00]. This DC power is boosted by a booster circuit (DC to DC converter) **5** and then transformed by a DC/AC inverter **6** to AC power which has the same voltage, frequency, and phase as those of an external AC power supply [TFa09][WBH00]. The output of the converter **3** is supplied to the load through a connection box **7** [TFa09][WBH00].

An engine controller **8** coordinates the opening of the fuel valve **20** during the start-up sequence and also during a steady-state operation [TFa09][WBH00]. A booster controller **9** controls the booster **5** and thus the DC output voltage [TFa09][WBH00]. A pump **22** and a pump controller **21** lubricate and cool the generator **2** using oil.

The booster circuit **5** together with the inverter **6** are controlled by an inverter control unit **18** based on a microcomputer. This control unit consists of an input DC voltage detector **10** for detecting the DC output of the booster circuit **5** (i.e., a DC voltage applied to the inverter circuit **6**), a voltage/current detector **11** for detecting the voltage of an external AC power supply or an output current of the converter **3**, and a PID and PWM controller (**12** and **13**) to coordinate the operation of the inverter **6** in accordance with pulse width modulation [TFa09][WBH00]. The inverter control unit **18** contains a switch controller to coordinate the switches in here to open and close, a start controller **15** that coordinates the starting-up sequence of the power supply unit, a linkage controller **16** to coordinate a synchronous operation with an external AC power line through the connection box **7**, and finally an external shut-off detector **17** to detect if an external AC power supply unit has been shut off (in other words to detect if the gas MT generator is disconnected from the AC power supply and is in autonomous operation, due to a power failure) [TFa09][WBH00].

An operation controller **19** generates a start/stop signal for the engine controller **8** and for the converter control unit **18** to start or to stop the power supply unit **24** and, at the same time, performs a control for setting the AC power (current or voltage) and its frequency to be output from this power supply unit [TFa09][WBH00].

The modem **23** receives a synchronizing pulse and a shut-off detection synchronizing signal output from the linkage controller **16** and transmits it to the other power supply units through the connection box **7** and bus **28** (see Figure 5.5). In this figure [TFa09][WBH00], the modem **23** receives at the same time the signals transmitted from the multiple power unit controller **26** through the bus **28**, the modem **27**, and the connection box **7**.

The control signals transmitted from the multiple power unit controller **26** through the modem **27** and bus **28** include a starting/stopping signal of the power supply units **25**, **29**, and **30** and set the values for the frequency, power, and other parameters that are output from **25**, **29**, and **30** [TFa09][WBH00].

An appropriate communicative environment means changing the bus **28** with suitable communication equipment such as, for example, a fiber optics, digital bus, or wireless installation [TFa09][WBH00].

The multiple power unit controller **26** determines the number of power supply units to be operated on the base of the power needed by the load and generates starting or stopping signals for these MTs [TFa09][WBH00].

As shown in Figure 5.6, the power supply units **25, 29,** and **30** have different voltage/current characteristics [TFa09][WBH00]. Such different characteristics can be set by an operator from a control panel (not shown in any of these figures for the sake of simplicity). The characteristics can be set by using the PWM controller **13** by using the booster controller **9,** which controls the booster **5,** or by conducting both of these operations. The current/voltage characteristics acting as control signals can be supplied to units **25, 29,** and **30** by the multiple power unit controller **26.** As shown in Figure 5.6, different rated voltages $V_1$, $V_2$, $V_3$ ($V_1 > V_2 > V_3$) are supplied to these three units. When the currents exceed predetermined rated values $I_1$, $I_2$, $I_3$, the output voltages gradually decrease from the rated voltages $V_1$, $V_2$, and $V_3$ [TFa09][WBH00].

Taking as an example the unit **25,** the output voltage is constant at $V_1$ disregarding the magnitude of the load current $I$ when the load current $I$ is equal to or less than a current $I_1$ and is gradually decreased from $V_1$ as the load current $I$ exceeds $I_1$ [TFa09][WBH00]. In the same manner, considering

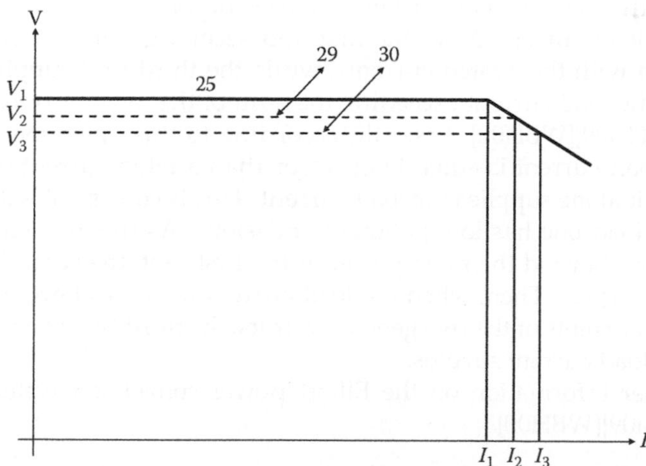


**FIGURE 5.6**
Basic principle of command for a cluster of three Calnetix units. (From T. Furuya et al., Electric power supply system. U.S. Patent 7,514,813, issued April 7, 2009, available online: http://www.uspto.gov [accessed March 22, 2013]; and W.B. Hall et al., Communications processor remote host and multiple unit control devices and methods for micropower generation systems. U.S. Patent 6,055,163, issued April 25, 2000, available online: http://www.uspto.gov [accessed March 22, 2013]. With permission.)

the power supply units **29** and **30**, while constant output voltages $V_2$ and $V_3$ are provided when the load currents correspond to $I_2$ and $I_3$, the respective output voltages are decreased from $V_2$ and $V_3$, as the load currents exceed $I_2$ and $I_3$.

To supply a load connected to the bus **28** with a current $I$ [TFa09][WBH00], when the power system starts, all the units begin to operate and generate voltages as a response to a start command coming from the multiple power unit controller **26**. Due to the fact that the rated voltages are set to $V_1 > V_2 > V_3$, the inverter **6** of the unit **25** outputs the highest voltage [TFa09][WBH00]. In such a way, the load current $I$ is supplied from the unit **25** in an initial phase when the load current is small, while the power supply units **29** and **30** hardly supply this load current.

As this current $I$ flowing through the bus **28** begins to increase to the value of $I_1$, the output voltage from the inverter **3** of the unit **25** gradually decreases from $V_1$ and becomes equal to the output voltage $V_2$ of the inverter **3** corresponding to the unit **29**. At the same time, the value of $I$ will increase to the current $I_2$. As load current exceeds the current $I_2$, the first unit **25** operates substantially at its rated parameters to supply $I_2$, and an overflow current $(I - I_2)$ flows from the second MT **29**. When a current exceeding $I_1$ is to flow from the unit **25**, the output voltage decreases, thus permitting the second unit **29** to supply a current. That is why the first generator supplies its rated current, and the second one supplies an overflow portion of the load current surpassing the rated current of the first power supply unit **25**.

Then, as the load current further increases beyond $I_3$, the third unit can supply the load current. Now, the first and second generators are supplying the load with their rated currents, while the third unit supplies only a portion of the load current exceeding the sum of the rated currents $2 \times I_3 = I - 2 \times I_3$ [TFa09][WBH00]. Basically, according to this operation principle, when the load current is equal to or lesser than a rated current of a single MT, this unit alone supplies the load current. This is consistent with the fact that at full load one has low pollutant emissions. As the load current further increases beyond the rated power of the first unit, the second unit will supply the surplus. Then, when the load current increases beyond the sum of the rated currents of the two generator units, the third turbogenerator will supply the load current surplus.

For further information on the Elliott power conversion system, please refer to [TFa09][WBH00].

# 6

# *Grid Measurements and General Features of a TA-100 Gas Microturbine*

The grid measurements in this chapter have been carried out on a former Elliott TA-100 (now called simply TA-100 and manufactured by Capstone) gas microturbine (MT) supplying electrical power to a national central network. Its capacity is valued at 100 kW. Figure 6.1 shows this type of MT.

Other important features of this turbogenerator are that it has the following [Cal-- ]:

- Built-in generator protection (this system protects the generator during unexpected faults)
- Weatherproof enclosure
- Low noise
- Fully integrated package (in the CHP version this includes power conditioning and controls, gas compressor, and heat recovery unit)
- Dual-mode controller (which allows the MT to act as a back-up generator for critical loads when combined with a UPS system, thus avoiding the expense of installing a classical stand-by generator)
- Remote monitoring

Other technical data are reported in Table 6.1.

The performances of the system on which the measurements described in the following pages have been carried out can be observed in Figure 6.2. The main features of this equipment are depicted in Figure 6.3. The measurements have been made using a power quality analyzer.

The waveforms of the voltages and currents as well as the phasor chart resulting from the grid measurements during the starting sequence of the MT can be seen in Figure 6.4. The D phase represents the null. In the moment depicted in the figure, the currents are distorted and have small values (around 2.5 A). The distortion appears to be due to the convertor of the MT oil pump which is among the first auxiliary systems to be turned on. In Figure 6.5 one can observe the same plots 3 minutes after start. In this case, the current values are bigger than the previous ones observed in Figure 6.4, having a value of approximately 20 A.

Due to the fact that the system is balanced, the voltage and current phasors of each phase should be coincident. Ten minutes after the starting sequence

**FIGURE 6.1**
General overview of the TA-100 gas MT. (Courtesy: Capstone Turbine Corporation.)

**TABLE 6.1**
Technical Data of a TA-100 Gas MT

| **Performance** | | **Thermal Output (hot water)** | | **Total Weight** | | **Batteries** |
|---|---|---|---|---|---|---|
| Electrical output | 107 kW | Flow | 3.8 l/s | Outdoor | 2050 kg | Two 12 volt, lead acid |
| Efficiency | 29.5% | | | | | |
| Maximum block loading | 100% | Water outlet temperature | 99°C | Indoor | 1860 kg | Parallel connected |
| Minimum load | 0 kW | System efficiency | >79% | | | |

*Source:* Adapted from Information on the Elliott (Calnetix) gas MT, http://www.tai-cepc.ro/fileadmin/download/pdf/Leafleats%20Capstone%20%26%20Calnetix/CalTA100.pdf (accessed April 29, 2013).

has been carried out (see Figure 6.6), one observes that the whole system is balanced (the current and voltage phasors for each phase overlap). The trend of the root mean square (RMS) value corresponding to the voltages and currents of the three phases can be seen in Figure 6.7.

The reason the curves in Figure 6.7 appear as bands and not as solid lines is that in this case the measurement of uncertainty has been taken into consideration. In other words, the upper and the lower boundaries of these bands impose the upper and lower values for measurement uncertainty.

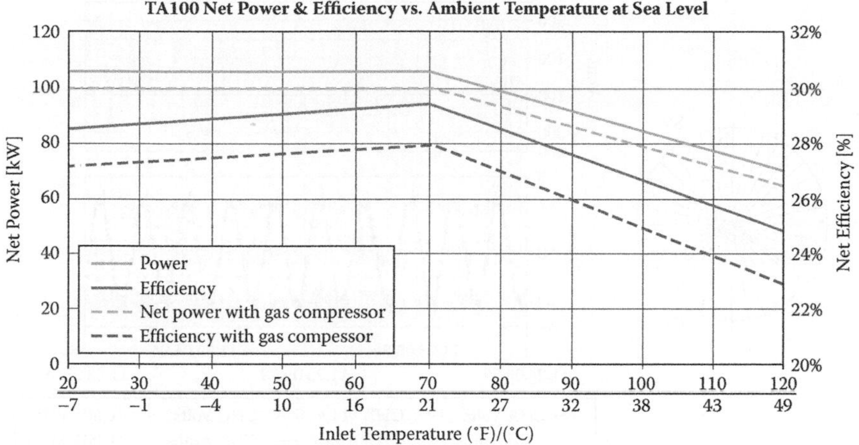


**FIGURE 6.2 (See color insert.)**
Net power and net efficiency as a function of temperature for a TA-100 natural gas MT. (Courtesy: Capstone Turbine Corporation.)

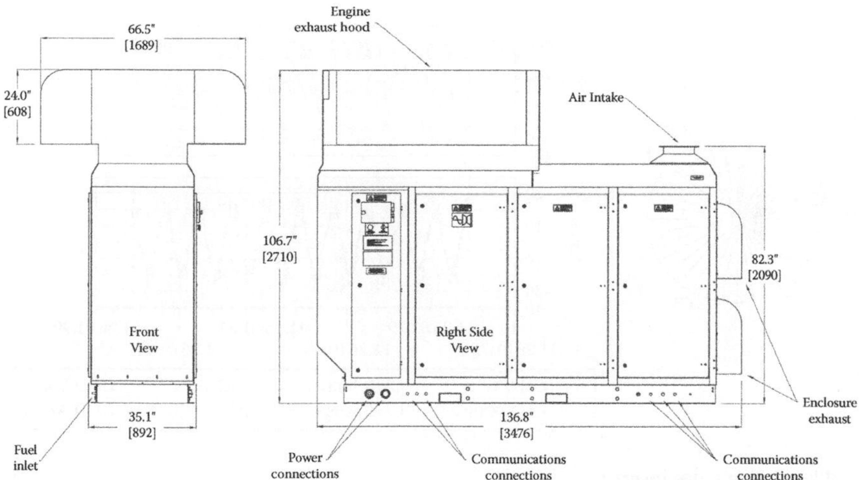


**FIGURE 6.3**
Main features of the TA-100 gas MT. (Courtesy: Capstone Turbine Corporation.)

The peak value of the voltage is reached approximately 1 minute after the start (see Figure 6.8). The current peak value is reached 2 minutes after start.

At the same time, as can be observed in Figure 6.9, the MT consumes power from the grid during approximately the first 8 minutes of the starting sequence in order to motor the unit. After these 8 minutes, the machine begins to deliver power back to the grid. For further details related to this, please refer to Appendix 1.

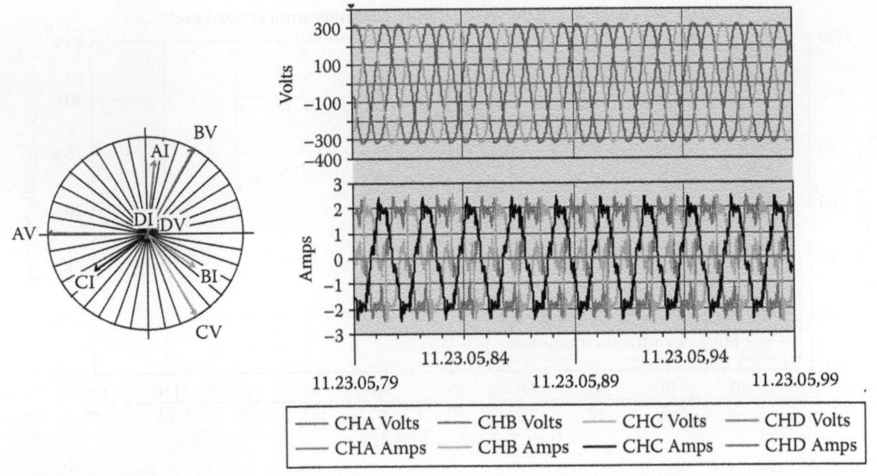


**FIGURE 6.4 (See color insert.)**
Approximate representation of the voltage, current waveforms and phasor chart during the MT starting sequence.

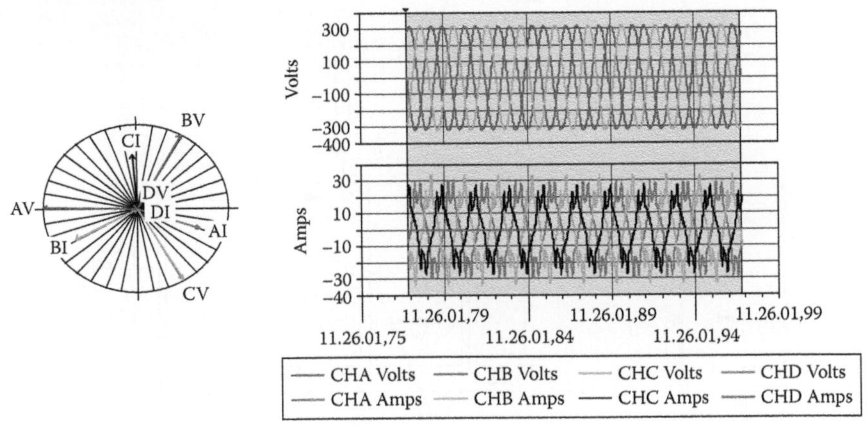


**FIGURE 6.5 (See color insert.)**
Approximate representation of the voltage, current waveforms and phasor chart 3 minutes after start.

The voltage and current unbalance (or RMS deviation) is at maximum after approximately 8 minutes with respect to the MT ignition moment (see Figures 6.10 and 6.12). The maximum voltage over deviation from the RMS value appears in the interval 11.31–11.39. The maximum phase-to-phase voltage over deviation from the RMS value appears instead in the interval 11.36–11.38 (see Figure 6.11).

Figure 6.13 indicates once more the fact that the MT begins to produce energy approximately 8 minutes after start. The maximum active and

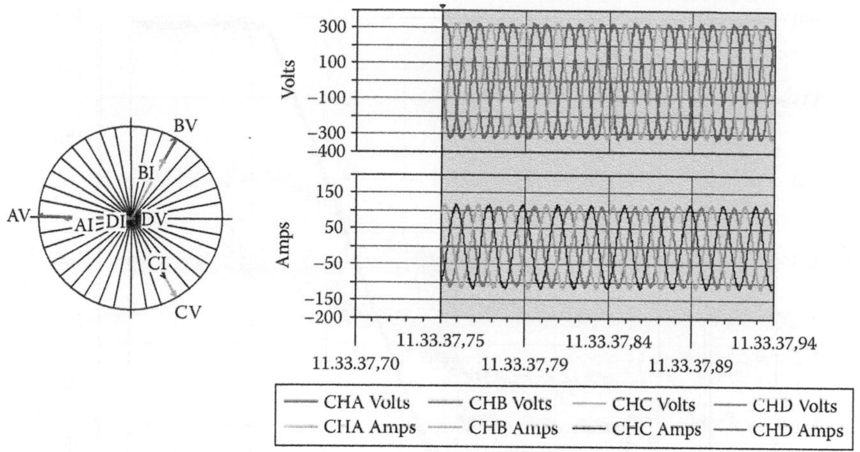


**FIGURE 6.6 (See color insert.)**
Approximate representation of the voltage, current waveforms and phasor chart 10 minutes after start.

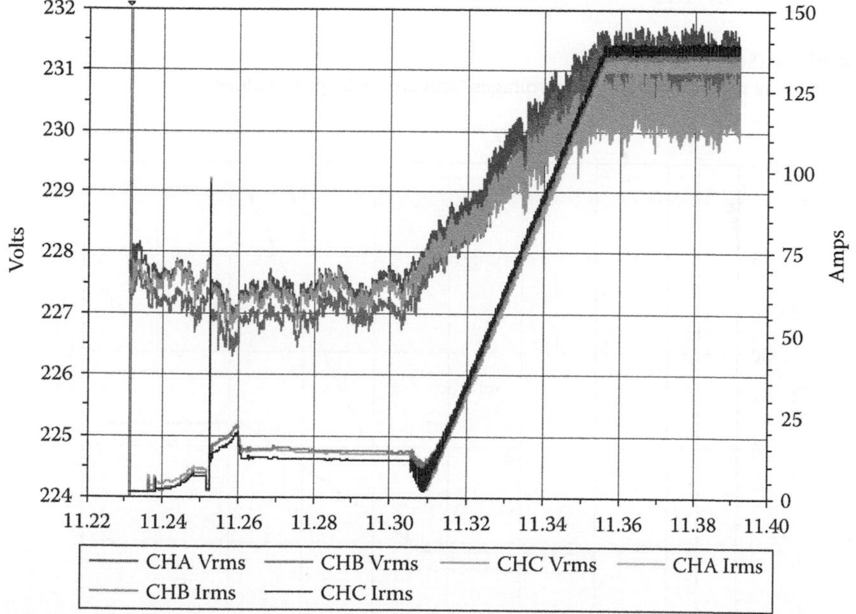


**FIGURE 6.7 (See color insert.)**
Approximate representation of the voltages' and currents' RMS trend.

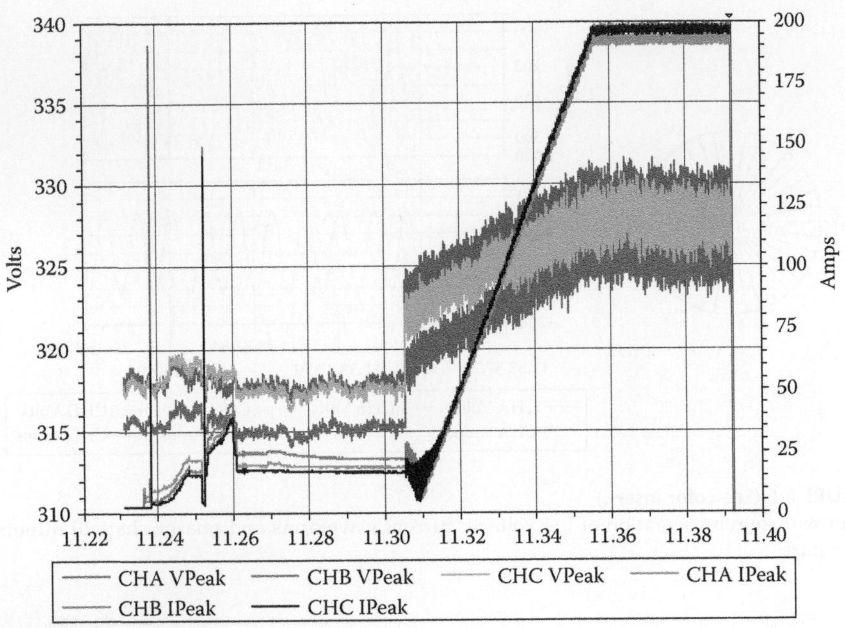


**FIGURE 6.8 (See color insert.)**
Approximate representation of the voltages' and currents' peak values.

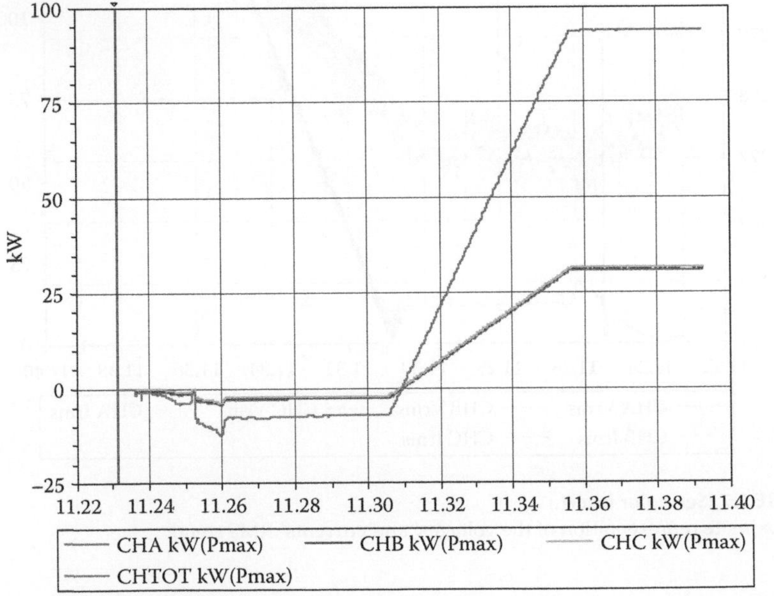


**FIGURE 6.9 (See color insert.)**
Approximate representation of the power peak trend.

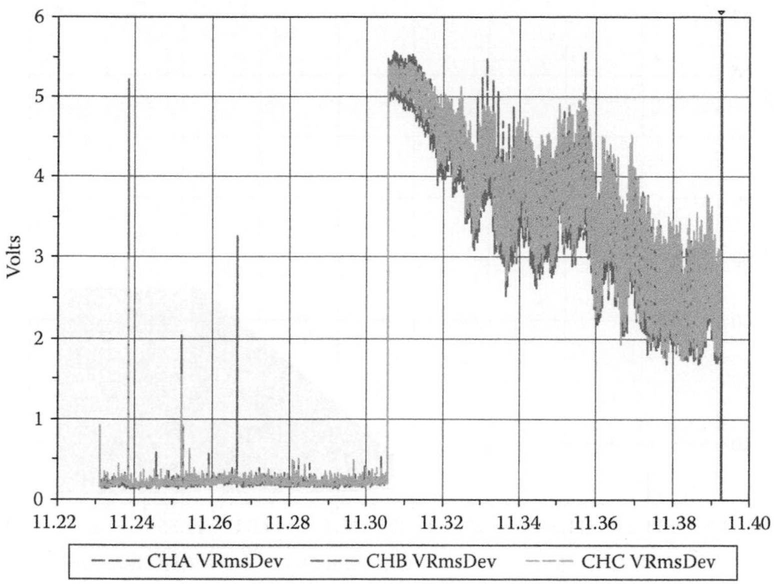


**FIGURE 6.10 (See color insert.)**
Approximate representation of the voltage unbalance.

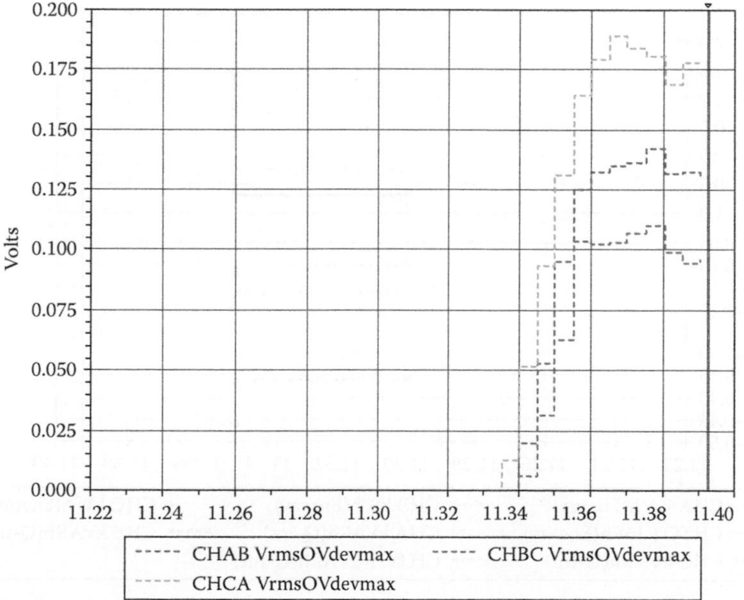


**FIGURE 6.11 (See color insert.)**
Approximate representation of the maximum over deviation between the phase-to-phase voltages.

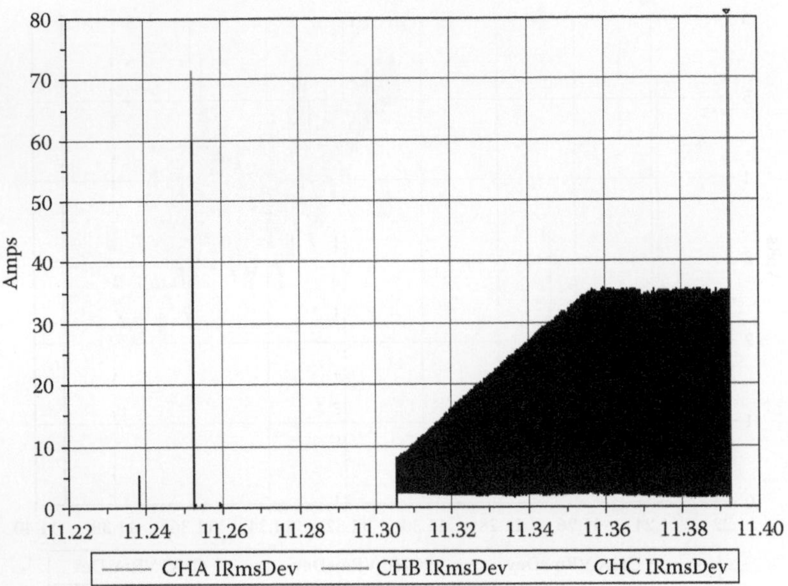


**FIGURE 6.12 (See color insert.)**
Approximate representation of the current unbalance.

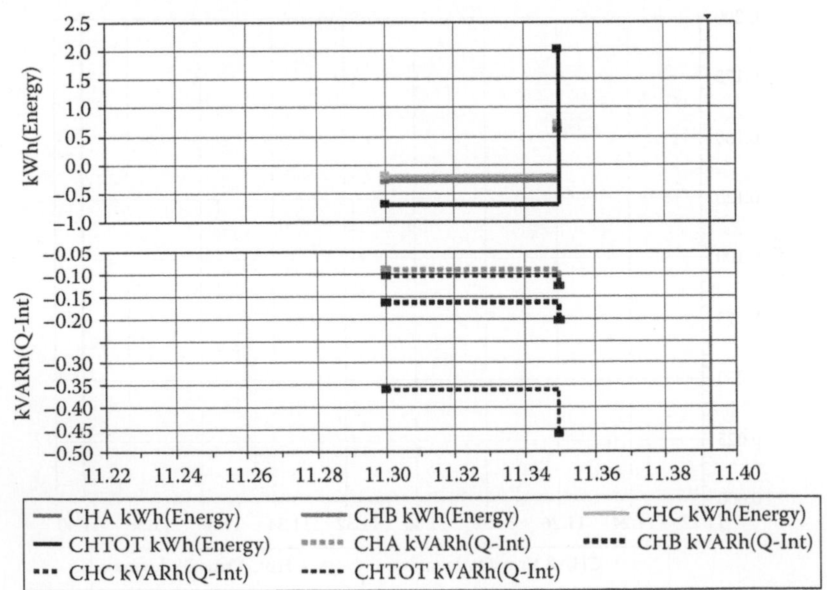


**FIGURE 6.13 (See color insert.)**
Approximate representation of the integrated active and reactive energy.

reactive energy is attained after 12 minutes compared to the machine ignition moment (see also Figure 6.9).

An important parameter that gives a strong indication whether the system is balanced or not is the displacement power factor (DPF). This represents the cosine between the voltage and the current phasors. In most cases the loads where one can encounter a poor DPF are [Kne99][M-- ][YCM94]: induction motors, transformers, reactive ballasts used for lighting and voltage control, and welding systems (non-inverter based) [LCH06][Smi02].

In other words, the bigger the DPF, the smaller would be the angle between the voltage and the current, thus meaning a balanced system [VBS08]. As one can observe from Figure 6.14, the DPF is smaller at 11.31 when the MT starts to deliver energy to the grid, and at 11.34 becomes maximum (when the unit produces more than 50% of the total required load, see also Figure 6.9). This would be yet more proof that the MT becomes more efficient when it functions at a minimum 50% of the rated power (see also Chapter 3).

Figure 6.15 gives a precise indication of the current needed by the MT in the ignition moment and also during the whole starting sequence. In regard to the voltages, the variations are not so important. Figure 6.16 illustrates the maximum positive, negative, and zero sequence components of the voltage, which in this case overlap. Figure 6.17 gives an important indication of the time interval when the voltage average harmonic and the voltage average interharmonic distortion, normalized to the fundamental, appear.

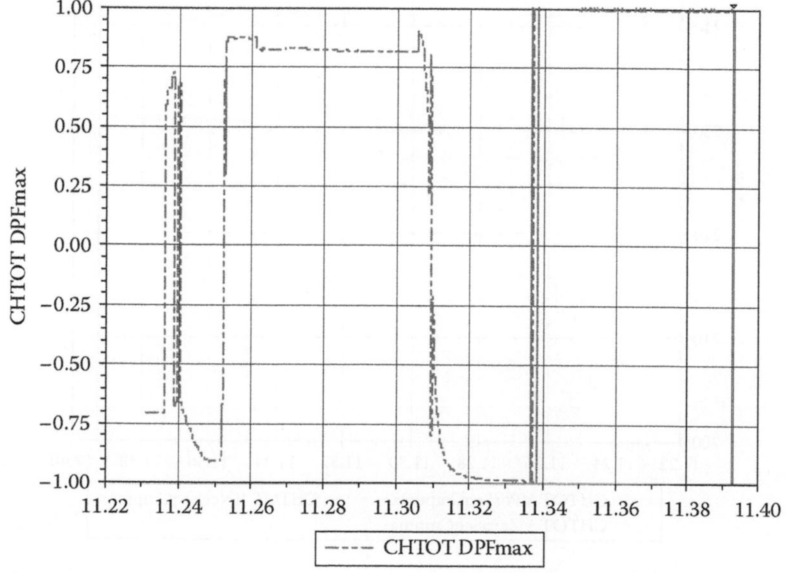


**FIGURE 6.14**
Approximate representation of the displacement power factor.

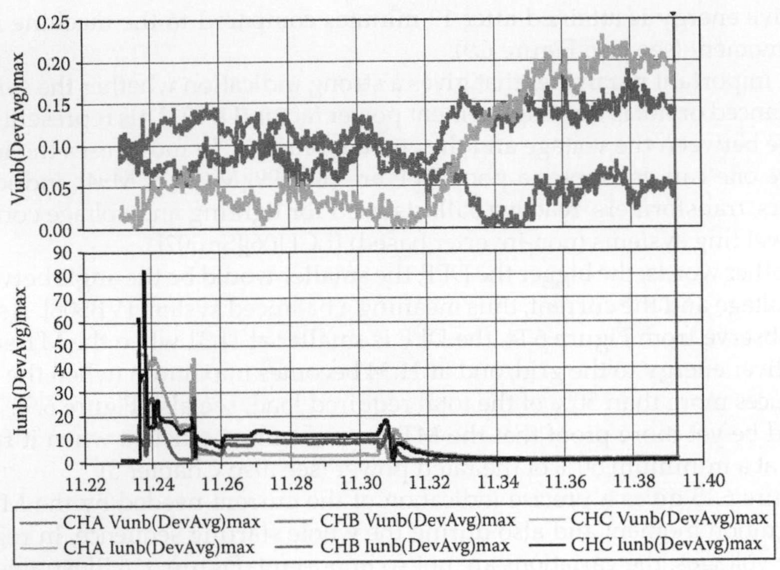


**FIGURE 6.15 (See color insert.)**
Approximate representation of the voltage and current unbalance (in%) according to the criterion of maximum deviation from the average.

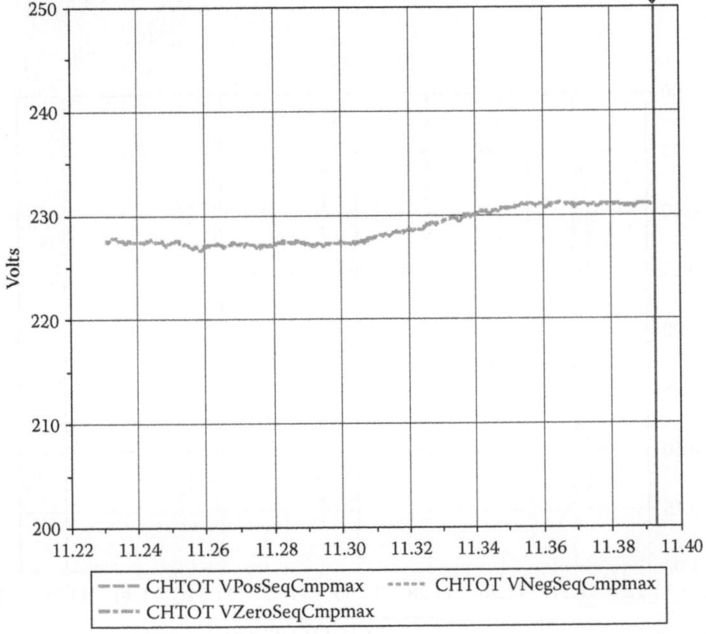


**FIGURE 6.16**
Approximate representation of the maximum positive, negative, and zero sequence components of voltage.

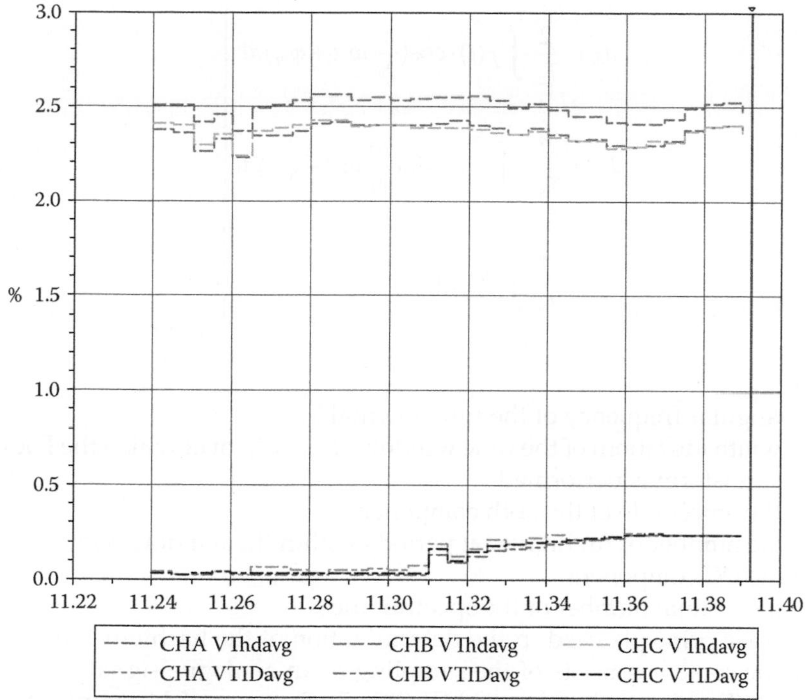


**FIGURE 6.17 (See color insert.)**
Approximate representation of the average voltage harmonic and average voltage interharmonic distortion (fundamental normalized).

The method used here to determine the harmonic and interharmonic content was the FFT (Fast Fourier transform) [AW04][IHM03][MG98]. This represents an improved version of the DFT (Discrete Fourier transform) [IEC02][NG97][VB09]:

$$f(t) = c_0 + \sum_{m=1}^{\infty} c_m \cdot \sin(\frac{m}{N}\omega_1 t + \phi_m) \tag{6.1}$$

$$\left.\begin{aligned} c_m &= | b_m + j a_m | = \sqrt{a_m^2 + b_m^2} \\ \phi_m &= \arctan(\frac{a_m}{b_m}) \quad if\ b_m \geq 0 \\ \phi_m &= \pi + \arctan(\frac{a_m}{b_m}) \quad if\ b_m < 0 \end{aligned}\right\} \tag{6.2}$$

$$\left.\begin{aligned} a_m &= \frac{2}{T_w}\int_0^{T_w} f(t)\cdot\cos\left(\frac{m}{N}\omega_1 t+\varphi_m\right)dt \\ b_m &= \frac{2}{T_w}\cdot\int_0^{T_w} f(t)\cdot\sin\left(\frac{m}{N}\omega_1 t+\varphi_m\right)dt \\ c_m &= \frac{1}{T_w}\int_0^{T_w} f(t)dt \end{aligned}\right\} \tag{6.3}$$

where:

$\omega_1$ = angular frequency of the fundamental

$T_w$ = width (duration) of the time window ($T_w = NT_1$) over which the Fourier transform is performed

$c_m$ = the amplitude of the $m$-th component

$N$ = the number of fundamental periods within the window width

$c_0$ = the DC component

$m$ = the ordinal number of the spectral line

The basic principle used in the determination of the harmonic and interharmonic content consists of the sampling of an analogue signal $f(t)$ which is then A/D converted and stored [IEC02]. Each group of $M$ samples forms a time window on which the DFT is performed [CL95]. The window width $T_w$ determines the frequency resolution $f_w = 1/T_w$ (the frequency separation of the spectral lines) for the analysis and thus the frequency basis for the result of the transform [CC06][TON00][ZT08]. That is why the window width $T_w$ must always be an integer of the fundamental period $T_1$ of the system voltage [QZ06]. The difference between the DFT and the FFT used in this case would be that the latter allows short computation times [TKN98] and requires at the same time that the number $M$ of samples is an integer power of 2 ($M = 2^i$, with $i \geq 10$, for instance) [IEC02].

The THD (total harmonic distortion) described below represents the ratio between the sum of all RMS values corresponding to all harmonic components $G_n$ up to a specified order ($H$) and the RMS value of the fundamental ($G_1$) [IEC02]:

$$THD = \sqrt{\sum_{n=2}^{H}\left(\frac{G_n}{G_1}\right)^2} \tag{6.4}$$

The TID (total interharmonic distortion) is determined based on the RMS value of an interharmonic component which represents the value of a spectral component with a frequency between two consecutive harmonic frequencies [Wat05].

Figure 6.18 gives an important indication of the time interval when the current average harmonic and the current average interharmonic distortion, normalized to the fundamental, appears. Unfortunately, these parameters have huge values (especially the interharmonic content) compared to the voltage.

In Figure 6.19, one can see another important parameter—the total phase harmonic power. Of course this indicator would have had smaller values if it were not for the values of the current harmonic and interharmonic distortions. This is confirmed in Figure 6.20.

The total harmonic content reaches its maximum (approximately 55% of the fundamental, see Figure 6.20) when the power flow changes its "direction"—in the moment when the MT ceases to consume electricity from the main network and begins to supply it to the grid. This moment is represented in the chart as a short transient phenomenon and should be disregarded. For clarification of this, one can study the contribution of even and odd current harmonics in the MT ignition moment (see Figures 6.21, 6.22, and 6.23).

For the current, it seems that the most important contribution to the total distortion, at the ignition moment, is brought by the harmonics between 2 and 20. As far as the voltages are concerned (see Figures 6.24, 6.25, and 6.26), the harmonic content is reduced. For the numerical data, see Appendix 2.

The harmonic trend as far as the current is concerned is different in the moment when the maximum shaft speed is reached (see Figures 6.27, 6.28,

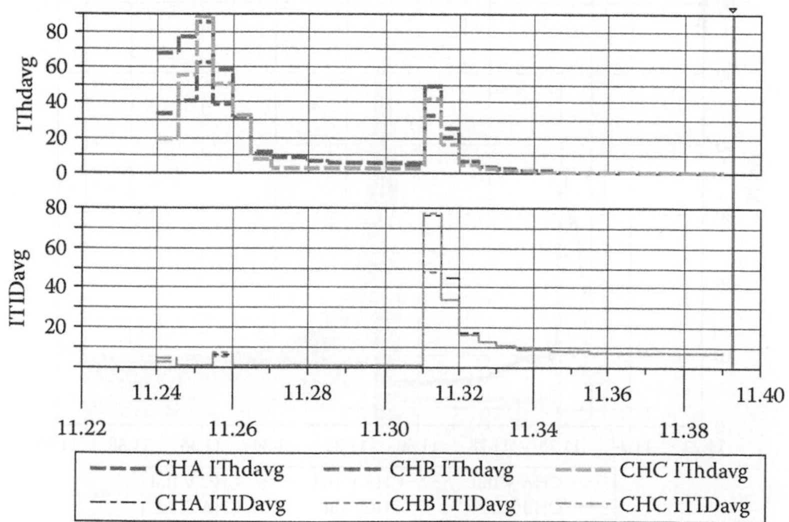


**FIGURE 6.18 (See color insert.)**
Approximate representation of the average current harmonic and average current interharmonic distortion (in%, fundamental normalized).

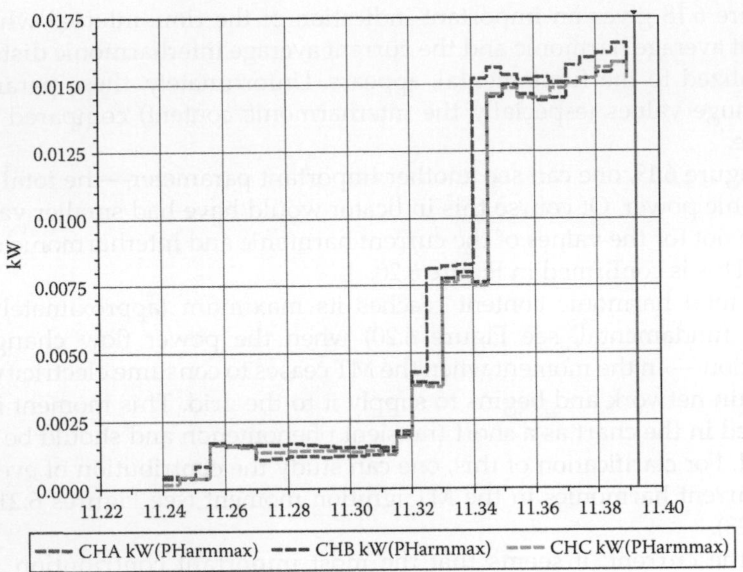


**FIGURE 6.19 (See color insert.)**
Approximate representation of the total phase harmonic power.

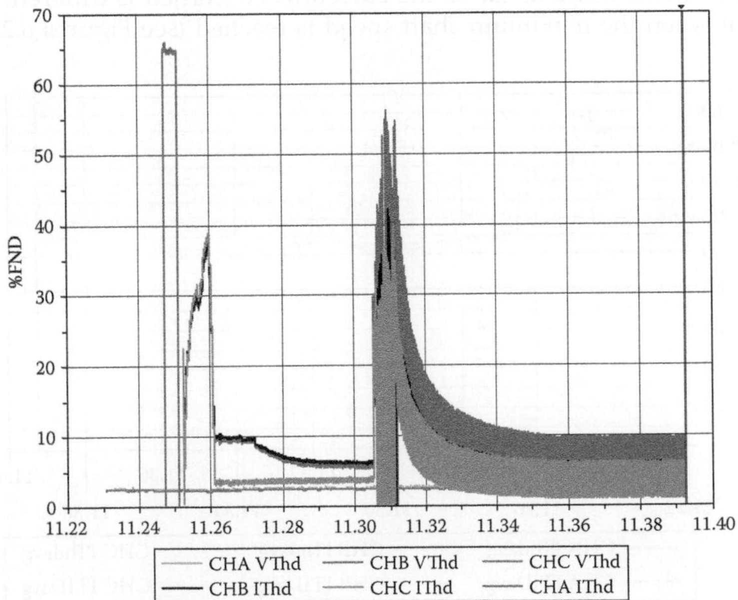


**FIGURE 6.20 (See color insert.)**
Approximate representation of the interdependency between the total harmonic content and the fundamental harmonic.

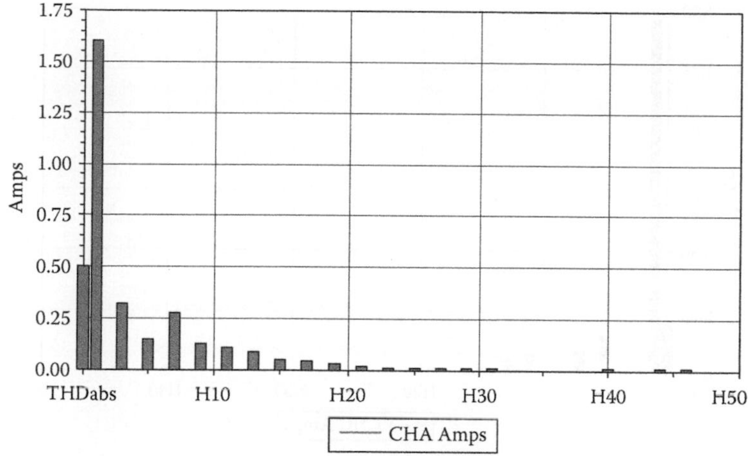


Total RMS: 1.68 Amps
DC Level : 0.01 Amps
Fundamental(H1) RMS: 1.60 Amps

| | |
|---|---|
| Total Harmonic Distortion (H02–H50): | 0.50 Amps RMS |
| Even contribution (H02–H50): | 0.02 Amps RMS |
| Odd contribution (H03–H49): | 0.50 Amps RMS |

**FIGURE 6.21**
Approximate representation of the contribution to the total distortion of the odd and even harmonics (phase A-current).

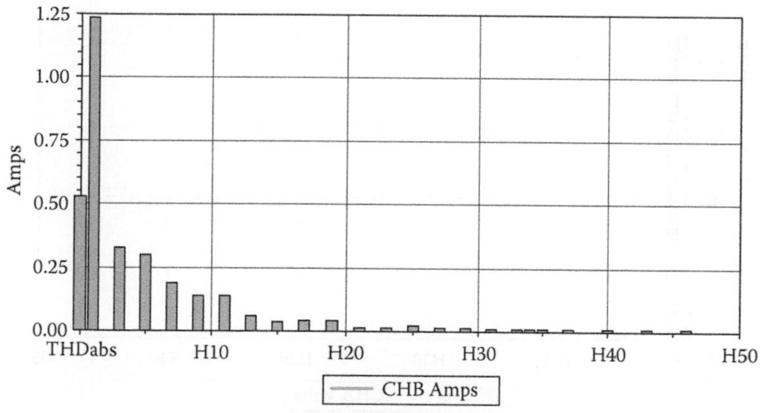


Total RMS: 1.35 Amps
DC Level : 0.01 Amps
Fundamental(H1) RMS: 1.23 Amps

| | |
|---|---|
| Total Harmonic Distortion (H02–H50): | 0.53 Amps RMS |
| Even contribution (H02–H50): | 0.02 Amps RMS |
| Odd contribution (H03–H49): | 0.53 Amps RMS |

**FIGURE 6.22**
Approximate representation of the contribution to the total distortion of the odd and even harmonics (phase B-current).

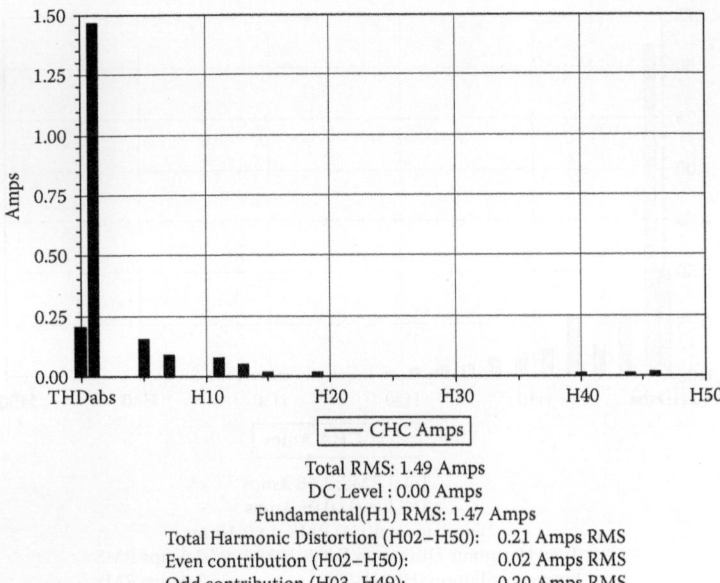


**FIGURE 6.23**
Approximate representation of the contribution to the total distortion of the odd and even harmonics (phase C-current).

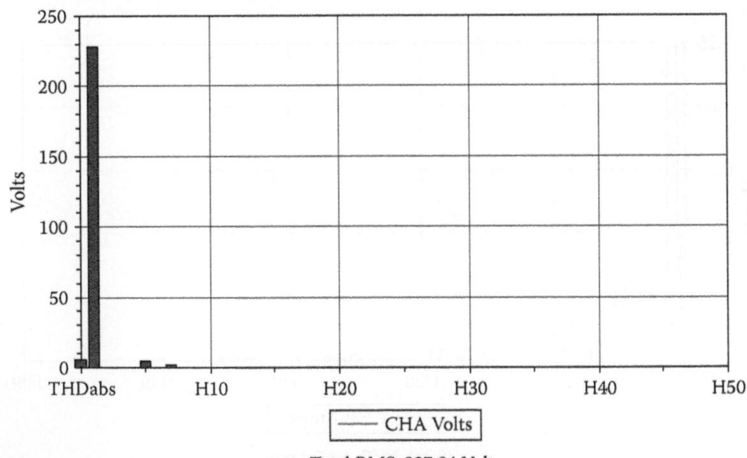


**FIGURE 6.24**
Approximate representation of the contribution to the total distortion of the odd and even harmonics (phase A-voltage).

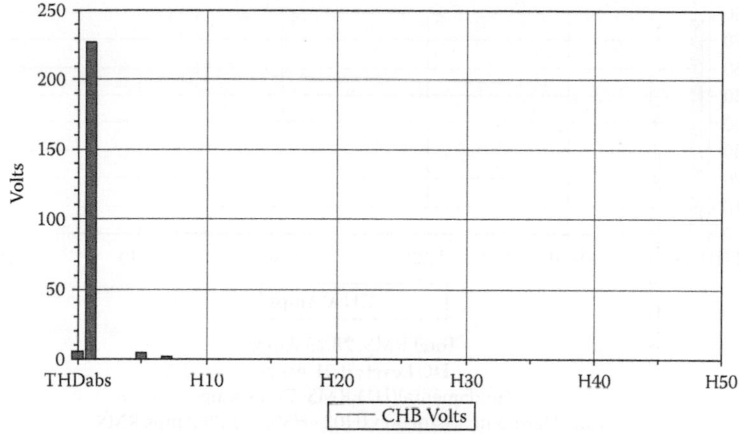


**FIGURE 6.25**
Approximate representation of the contribution to the total distortion of the odd and even harmonics (phase B-voltage).

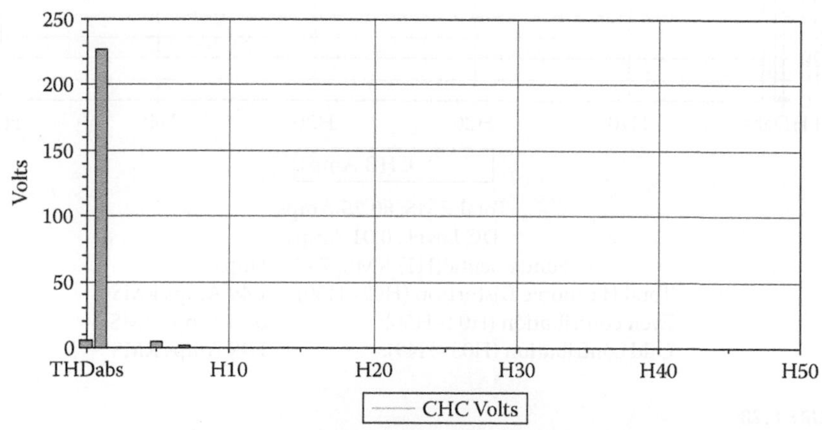


**FIGURE 6.26**
Approximate representation of the contribution to the total distortion of the odd and even harmonics (phase C-voltage).

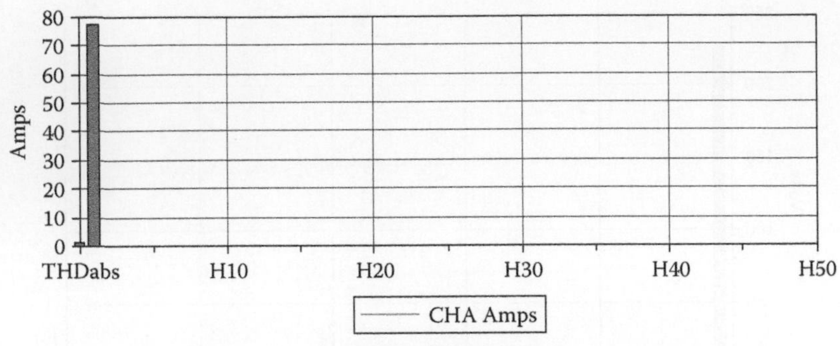


Total RMS: 78.23 Amps
DC Level : 0.01 Amps
Fundamental(H1) RMS: 77.55 Amps

| | |
|---|---|
| Total Harmonic Distortion (H02–H50): | 1.39 Amps RMS |
| Even contribution (H02–H50): | 0.12 Amps RMS |
| Odd contribution (H03–H49): | 1.39 Amps RMS |

**FIGURE 6.27**
Approximate representation of the contribution to the total distortion of the odd and even harmonics 10 minutes after ignition (phase A-current).

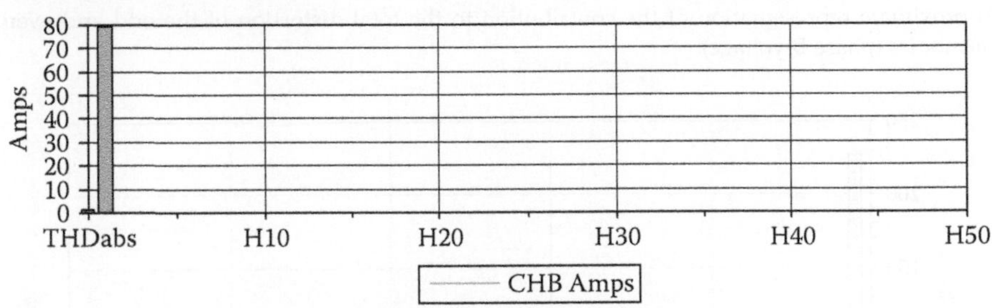


Total RMS: 80.25 Amps
DC Level : 0.01 Amps
Fundamental(H1) RMS: 79.55 Amps

| | |
|---|---|
| Total Harmonic Distortion (H02–H50): | 1.49 Amps RMS |
| Even contribution (H02–H50): | 0.14 Amps RMS |
| Odd contribution (H03–H49): | 1.49 Amps RMS |

**FIGURE 6.28**
Approximate representation of the contribution to the total distortion of the odd and even harmonics 10 minutes after ignition (phase B-current).

and 6.29), while the voltage harmonic trend is practically identical to the one in the ignition moment (see Figures 6.30, 6.31, and 6.32). For a broader picture of the harmonics 10 minutes after start, see Appendix 3.

Another interesting parameter is the time necessary to perform a load variation from 64 kW to 53 kW. As shown in Figures 6.33, 6.34, and 6.35, the time needed is approximately 14 sec. The voltage and current waveforms

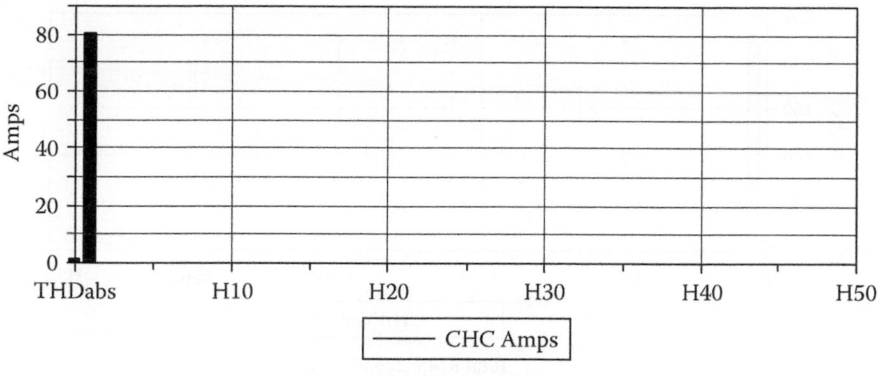


Total RMS: 80.99 Amps
DC Level : −0.01 Amps
Fundamental(H1) RMS: 80.26 Amps

| | |
|---|---|
| Total Harmonic Distortion (H02–H50): | 1.16 Amps RMS |
| Even contribution (H02–H50): | 0.19 Amps RMS |
| Odd contribution (H03–H49): | 1.15 Amps RMS |

**FIGURE 6.29**
Approximate representation of the contribution to the total distortion of the odd and even harmonics 10 minutes after ignition (phase C-current).

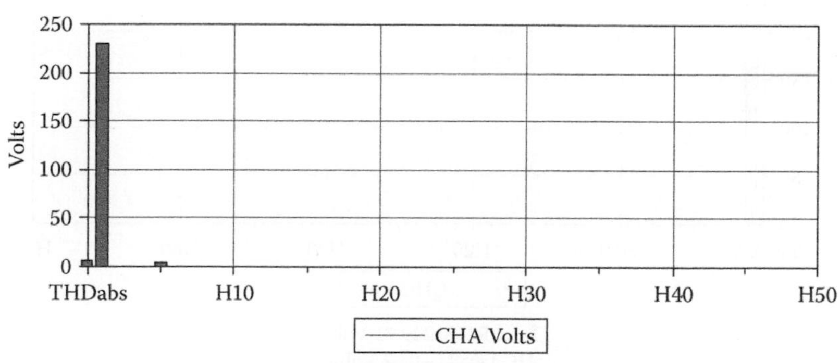


Total RMS: 230.30 Volts
DC Level : 0.00 Volts
Fundamental(H1) RMS: 230.20 Volts

| | |
|---|---|
| Total Harmonic Distortion (H02–H50): | 5.68 Volts RMS |
| Even contribution (H02–H50): | 0.16 Volts RMS |
| Odd contribution (H03–H49): | 5.68 Volts RMS |

**FIGURE 6.30**
Approximate representation of the contribution to the total distortion of the odd and even harmonics 10 minutes after ignition (phase A-voltage).

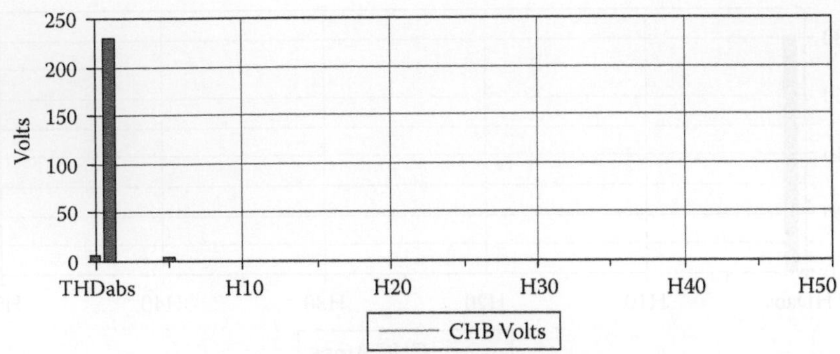


Total RMS: 229.97 Volts
DC Level : 0.00 Volts
Fundamental(H1) RMS: 229.88 Volts

| | |
|---|---|
| Total Harmonic Distortion (H02–H50): | 5.34 Volts RMS |
| Even contribution (H02–H50): | 0.21 Volts RMS |
| Odd contribution (H03–H49): | 5.34 Volts RMS |

**FIGURE 6.31**
Approximate representation of the contribution to the total distortion of the odd and even harmonics 10 minutes after ignition (phase B-voltage).

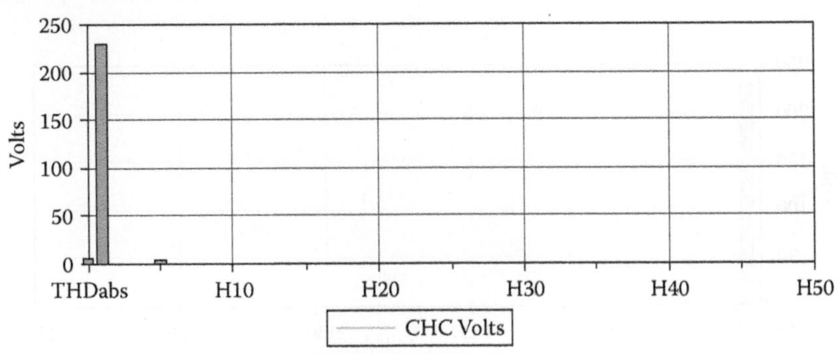


Total RMS: 229.71 Volts
DC Level : –0.02 Volts
Fundamental(H1) RMS: 229.62 Volts

| | |
|---|---|
| Total Harmonic Distortion (H02–H50): | 5.33 Volts RMS |
| Even contribution (H02–H50): | 0.19 Volts RMS |
| Odd contribution (H03–H49): | 5.33 Volts RMS |

**FIGURE 6.32**
Approximate representation of the contribution to the total distortion of the odd and even harmonics 10 minutes after ignition (phase C-voltage).

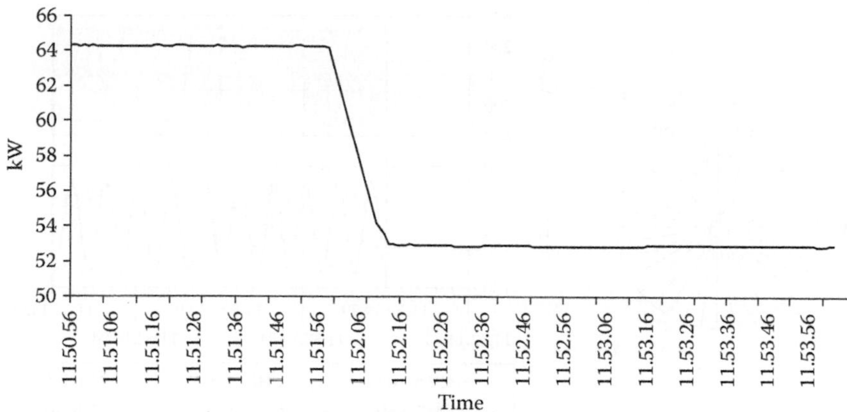


**FIGURE 6.33**
Approximate representation of the load variation from 64 kW to 53 kW.

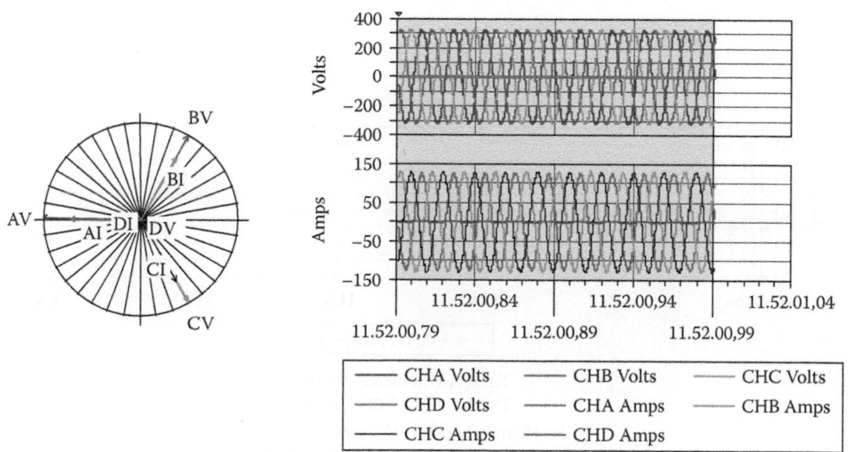


**FIGURE 6.34 (See color insert.)**
Approximate representation of the voltage and current waveforms prior to load variation.

corresponding to the two moments (before and after the load variation) can be seen in Figures 6.34 and 6.35. Another key element is that during this transition, the entire system remains balanced (see the phasor chart in Figures 6.34 and 6.35).

The harmonic and interharmonic content during the transition, on the other hand, is very low, especially when compared to the ignition sequence (not only from the point of view of the current but also from the voltage point of view, as shown in Figures 6.36 through 6.42).

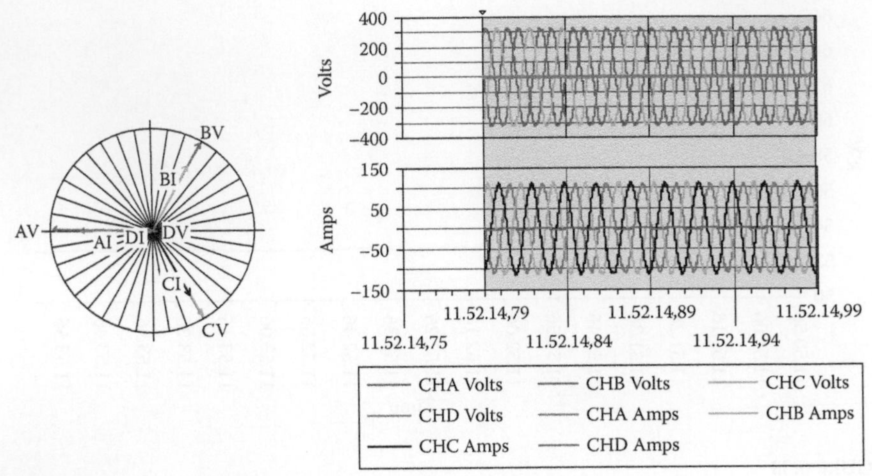


**FIGURE 6.35 (See color insert.)**
Approximate representation of the contribution and current waveforms after load variation.

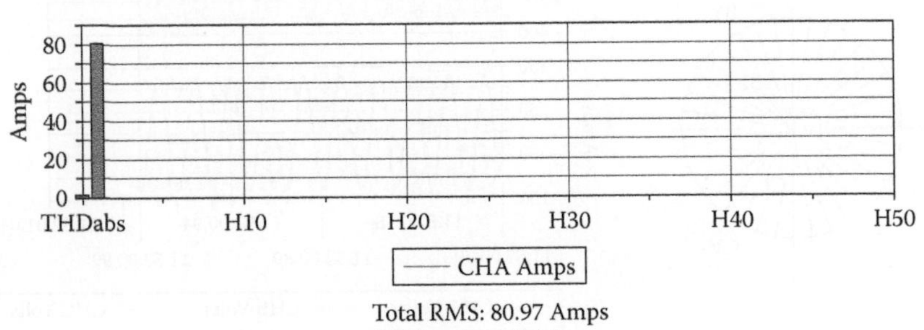


**FIGURE 6.36**
Approximate representation of the contribution to the total distortion of the odd and even harmonics during load variation (phase A-current).

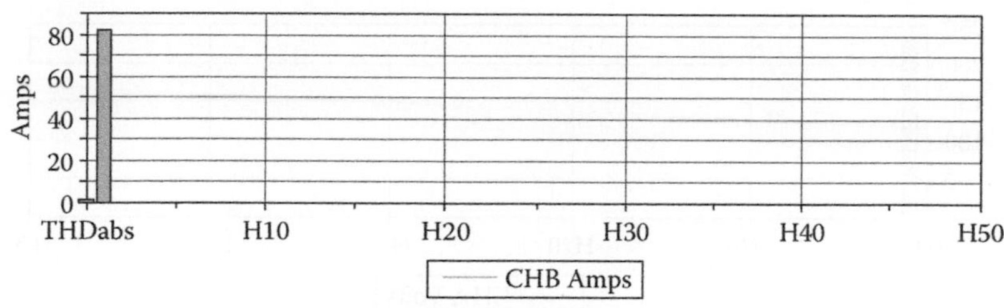


**FIGURE 6.37**
Approximate representation of the contribution to the total distortion of the odd and even harmonics during load variation (phase B-current).

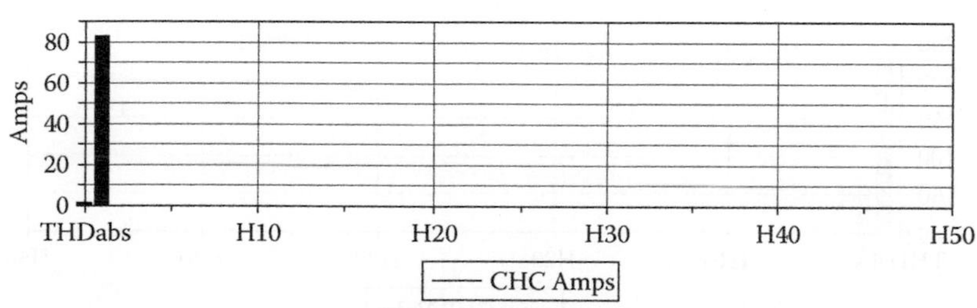


**FIGURE 6.38**
Approximate representation of the contribution to the total distortion of the odd and even harmonics during load variation (phase C-current).

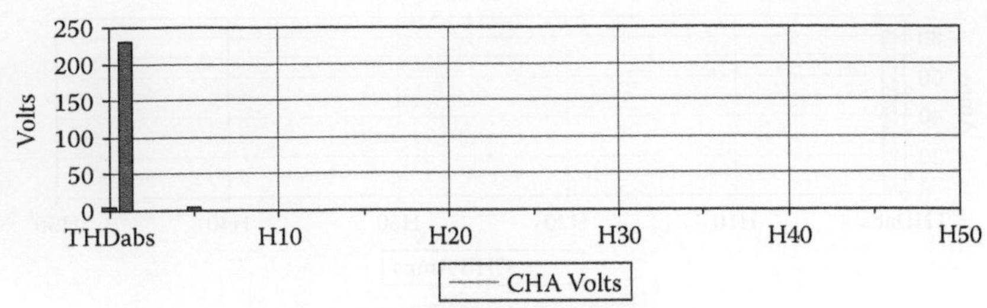


**FIGURE 6.39**
Approximate representation of the contribution to the total distortion of the odd and even harmonics during load variation (phase A-voltage).

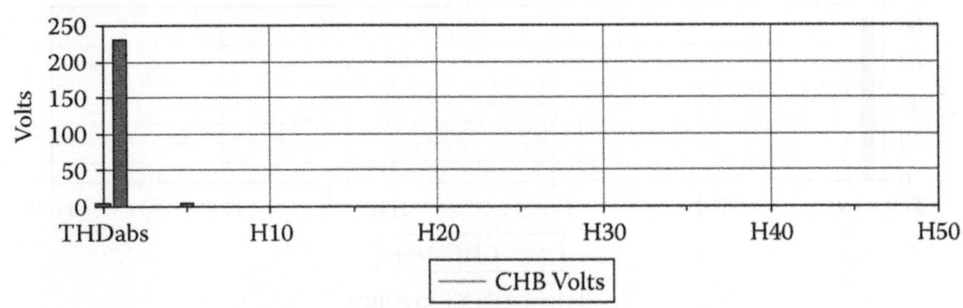


**FIGURE 6.40**
Approximate representation of the contribution to the total distortion of the odd and even harmonics during load variation (phase B-voltage).

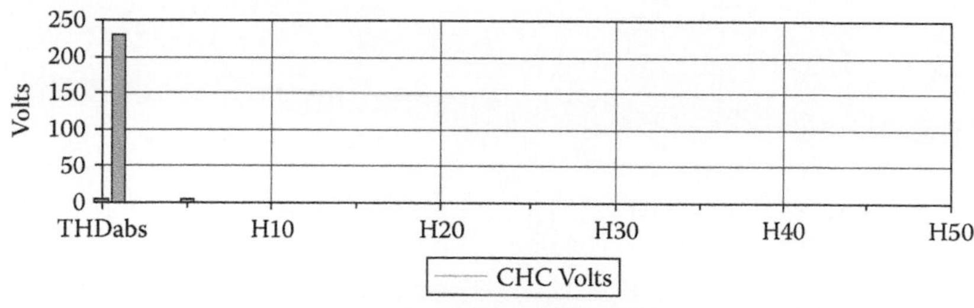


Total RMS: 229.59 Volts
DC Level : 0.00 Volts
Fundamental(H1) RMS: 229.51 Volts

| | |
|---|---|
| Total Harmonic Distortion (H02–H50): | 5.05 Volts RMS |
| Even contribution (H02–H50): | 0.34 Volts RMS |
| Odd contribution (H03–H49): | 5.04 Volts RMS |

**FIGURE 6.41**
Approximate representation of the contribution to the total distortion of odd and even harmonics during load variation (phase C-voltage).

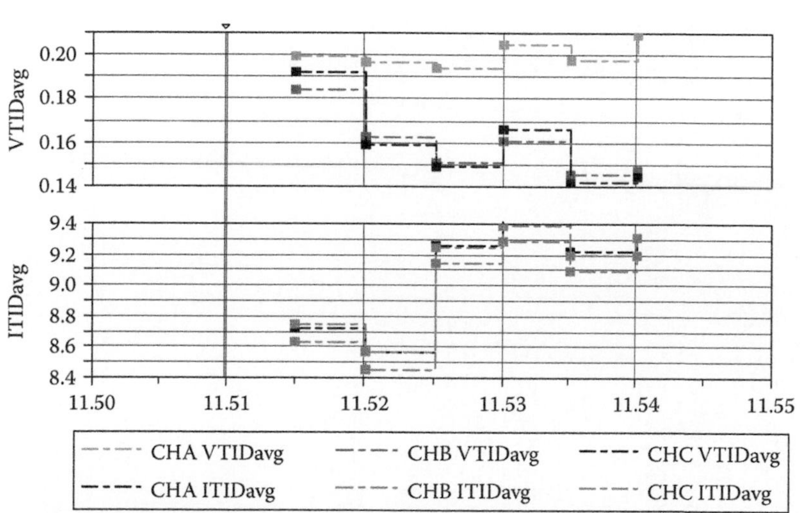


**FIGURE 6.42 (See color insert.)**
Approximate representation of the current average harmonic and current average interharmonic distortions (in%, fundamental normalized).

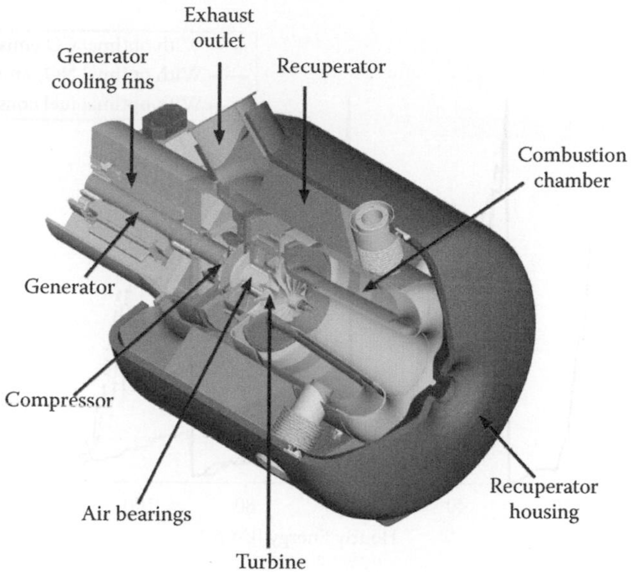


**FIGURE 2.3**
Cross section of a C30 natural gas MT. (Courtesy: Capstone Turbine Corporation.)

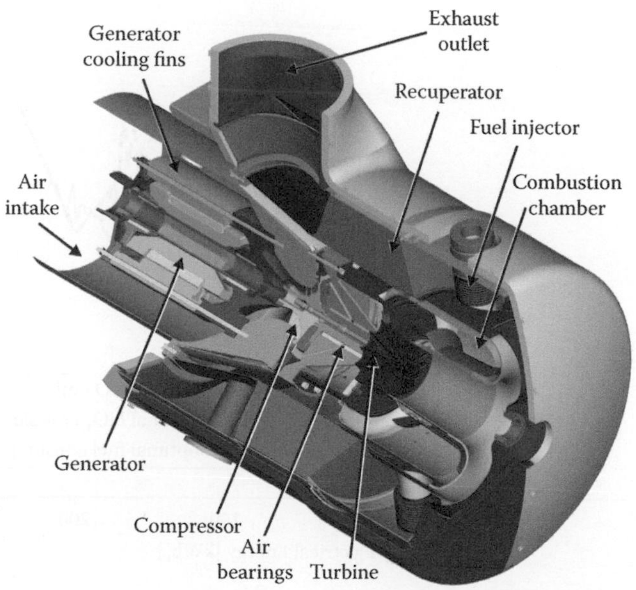


**FIGURE 2.4**
Cross section of a C65 natural gas MT. (Courtesy: Capstone Turbine Corporation.)

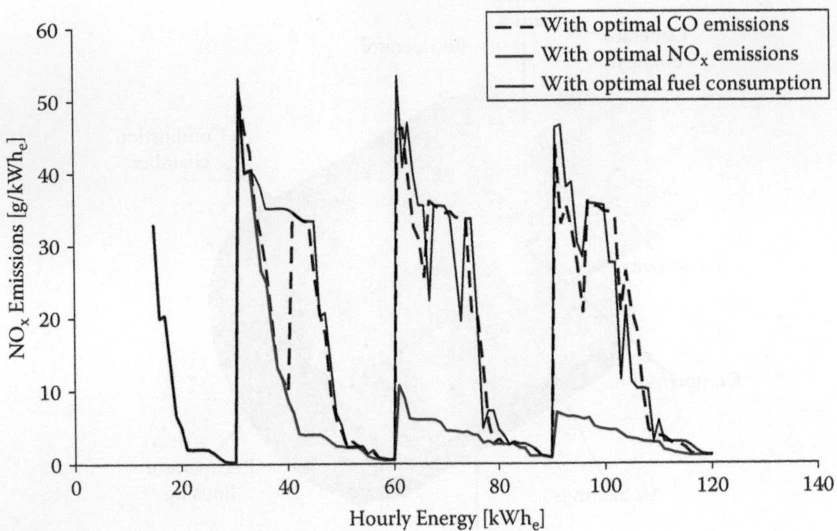


**FIGURE 3.5**
$NO_x$ emissions for different optimization scenarios in the case of a 30 kW$_e$ MT. (From A.V. Boicea, G. Chicco, and P. Mancarella, Optimal operation of a 30 kW natural gas MT cluster, *Buletinul Stiintific al Universitatii Politehnica Bucuresti, Seria C,* 73(1), 211–222, 2011. With permission.)

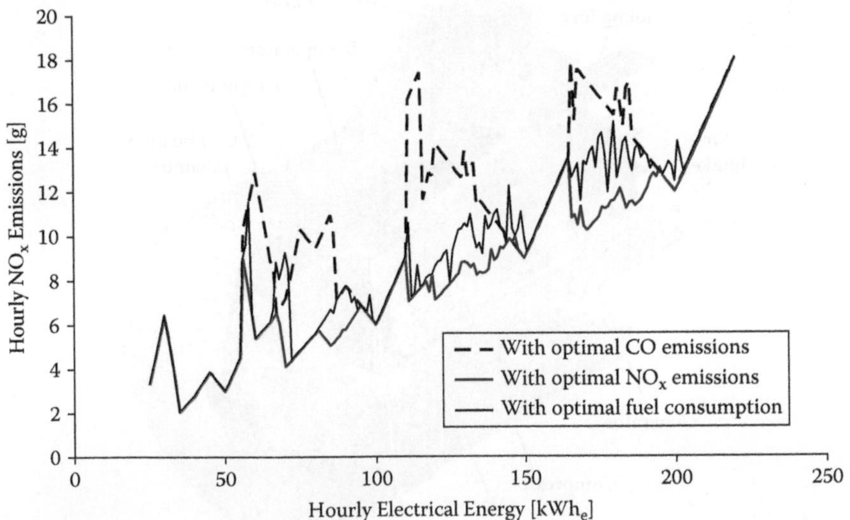


**FIGURE 3.6**
$NO_x$ emissions for different optimization scenarios in the case of a 60 kW$_e$ MT. (From A.V. Boicea, G. Chicco, and P. Mancarella, Optimal operation of a MT cluster with partial-load efficiency and emission characterization, Powertech, 2009 IEEE Bucharest, June 28–July 2, 2009, Pages 1–8. With permission.)

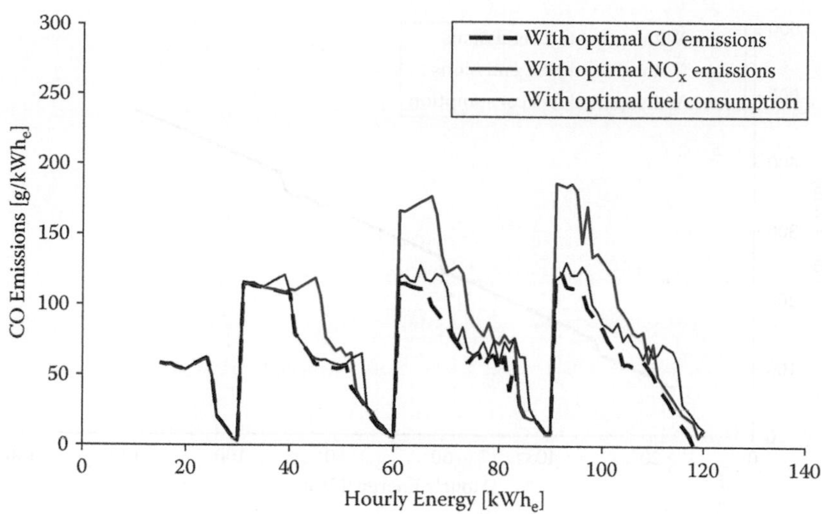


**FIGURE 3.7**
CO emissions for different optimization scenarios in the case of a 30 kW$_e$ MT. (From A.V. Boicea, G. Chicco, and P. Mancarella, Optimal operation of a 30 kW natural gas MT cluster, *Buletinul Stiintific al Universitatii Politehnica Bucuresti, Seria C*, 73(1), 211–222, 2011. With permission.)

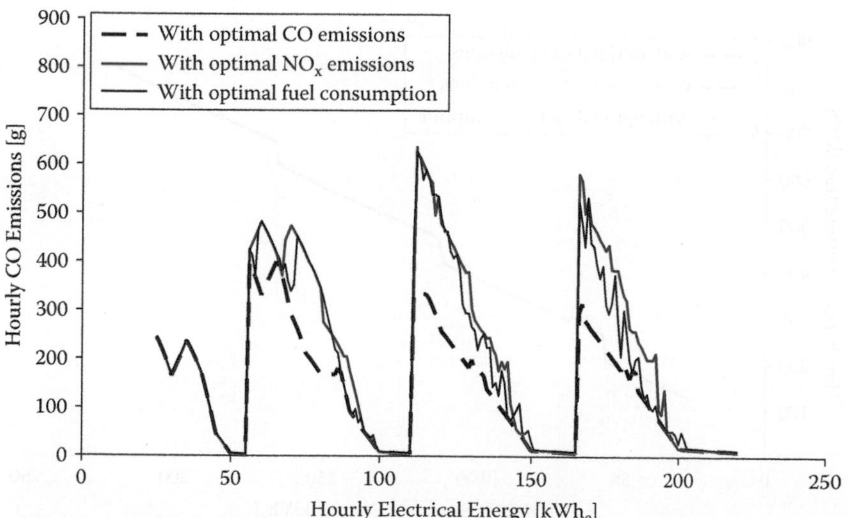


**FIGURE 3.8**
CO emissions for different optimization objectives in the case of a 60 kW$_e$ MT. (From A.V. Boicea, G. Chicco, and P. Mancarella, Optimal operation of a MT cluster with partial-load efficiency and emission characterization, Powertech, 2009 IEEE Bucharest, June 28–July 2, 2009, Pages 1–8. With permission.)

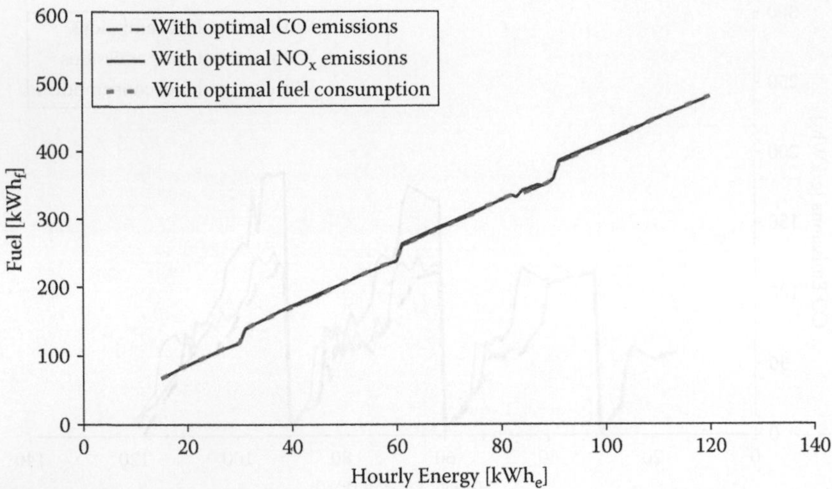


**FIGURE 3.9**
Fuel consumption for different optimization scenarios in the case of a 30 $kW_e$ MT. (From A.V. Boicea, G. Chicco, and P. Mancarella, Optimal operation of a 30 kW natural gas MT cluster, *Buletinul Stiintific al Universitatii Politehnica Bucuresti, Seria C*, 73(1), 211–222, 2011. With permission.)

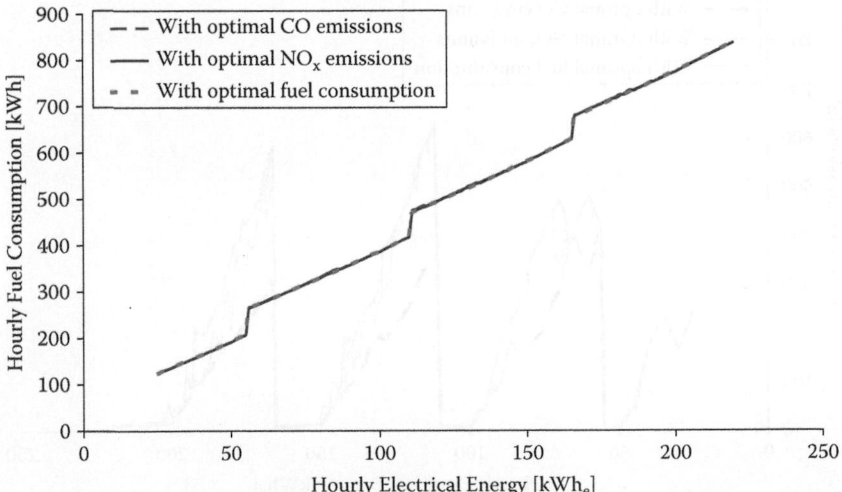


**FIGURE 3.10**
Fuel consumption for different optimization objectives in the case of a 60 $kW_e$ MT. (From A.V. Boicea, G. Chicco, and P. Mancarella, Optimal operation of a MT cluster with partial-load efficiency and emission characterization, Powertech, 2009 IEEE Bucharest, June 28–July 2, 2009, Pages 1–8. With permission.)

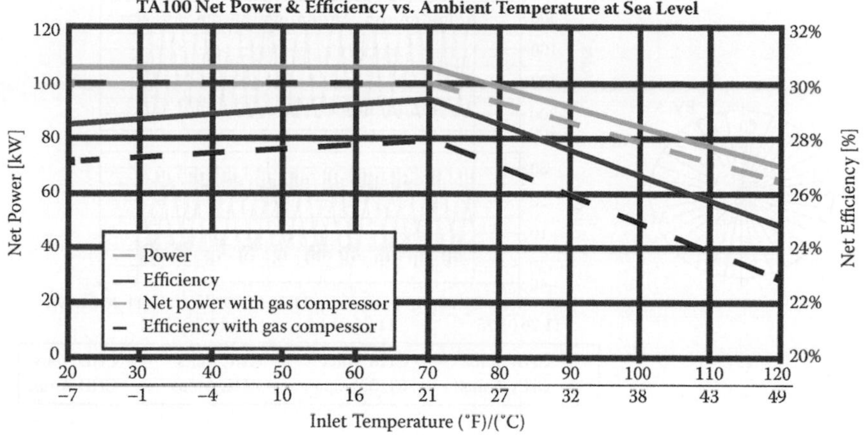


**FIGURE 6.2**
Net power and net efficiency as a function of temperature for a TA-100 natural gas MT. (Courtesy: Capstone Turbine Corporation.)

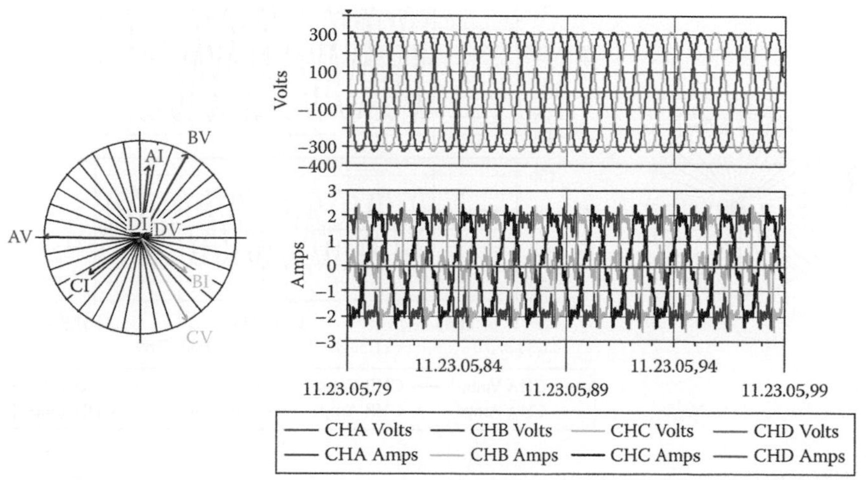


**FIGURE 6.4**
Approximate representation of the voltage, current waveforms and phasor chart during the MT starting sequence.

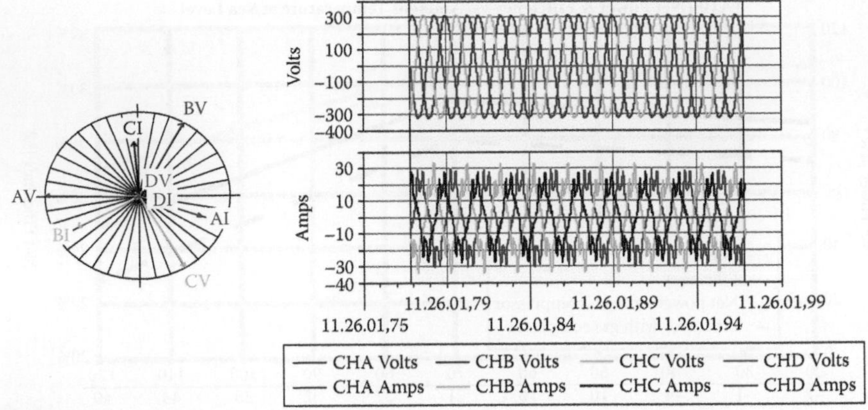


**FIGURE 6.5**
Approximate representation of the voltage, current waveforms and phasor chart 3 minutes after start.

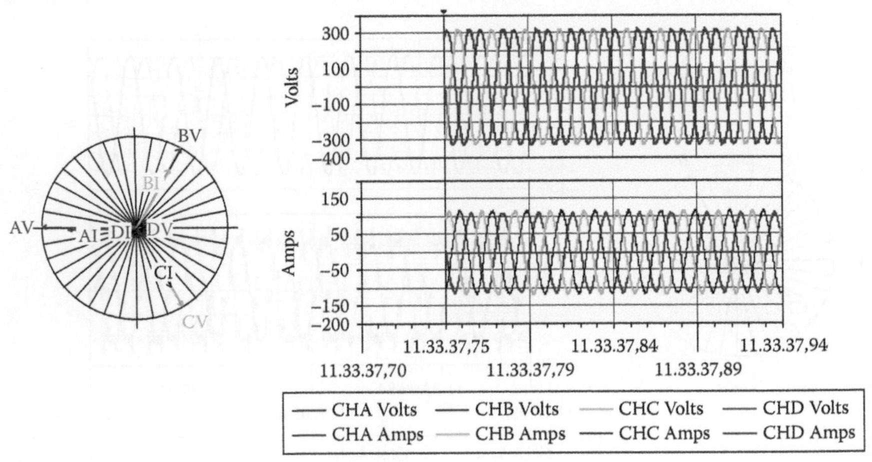


**FIGURE 6.6**
Approximate representation of the voltage, current waveforms and phasor chart 10 minutes after start.

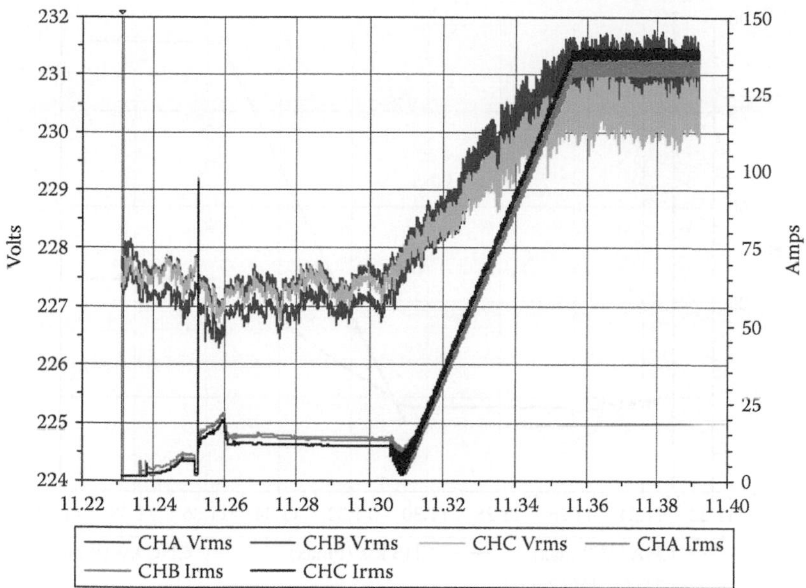


**FIGURE 6.7**
Approximate representation of the voltages' and currents' RMS trend.

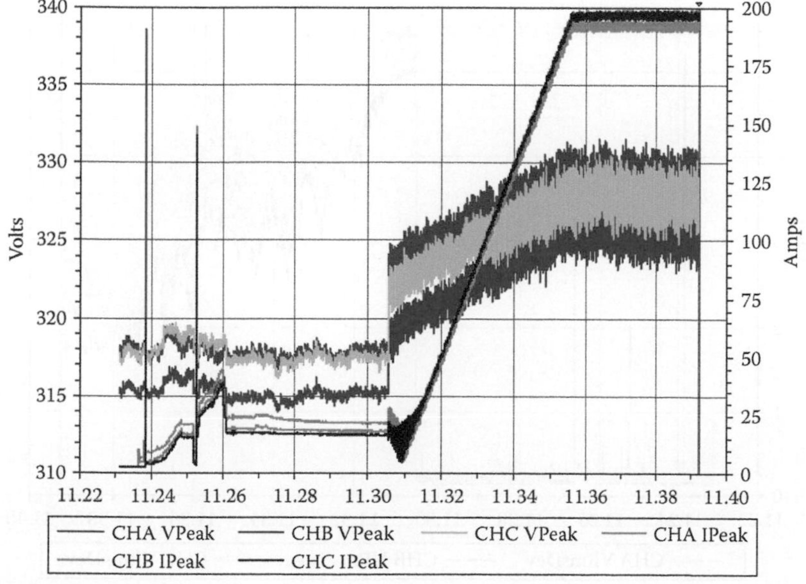


**FIGURE 6.8**
Approximate representation of the voltages' and currents' peak values.

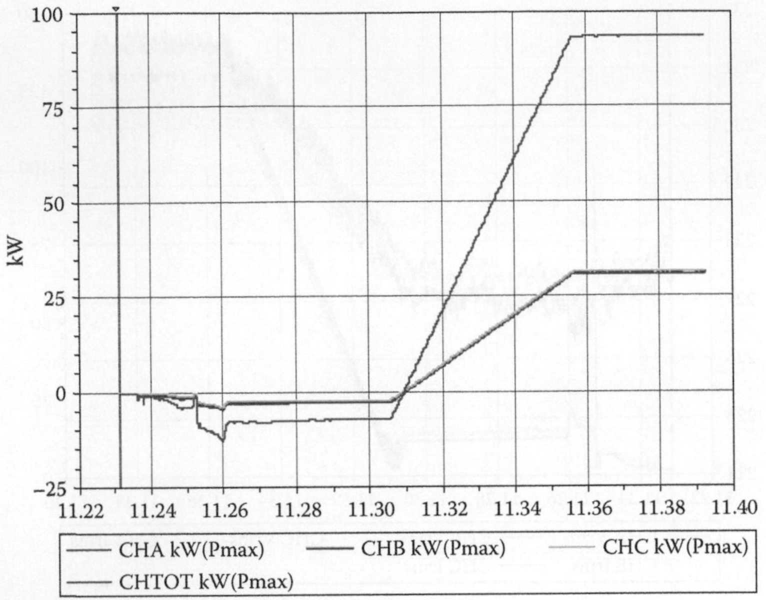


**FIGURE 6.9**
Approximate representation of the power peak trend.

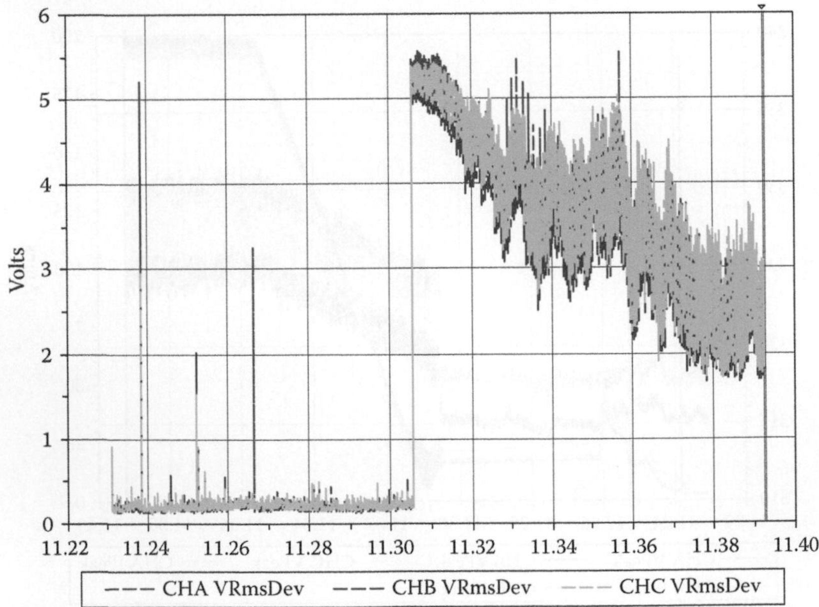


**FIGURE 6.10**
Approximate representation of the voltage unbalance.

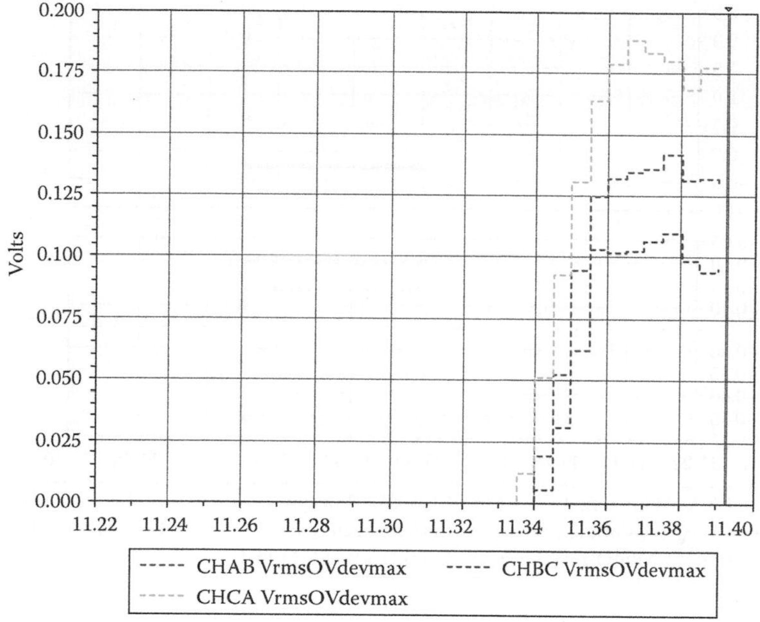


**FIGURE 6.11**
Approximate representation of the maximum over deviation between the phase-to-phase voltages.

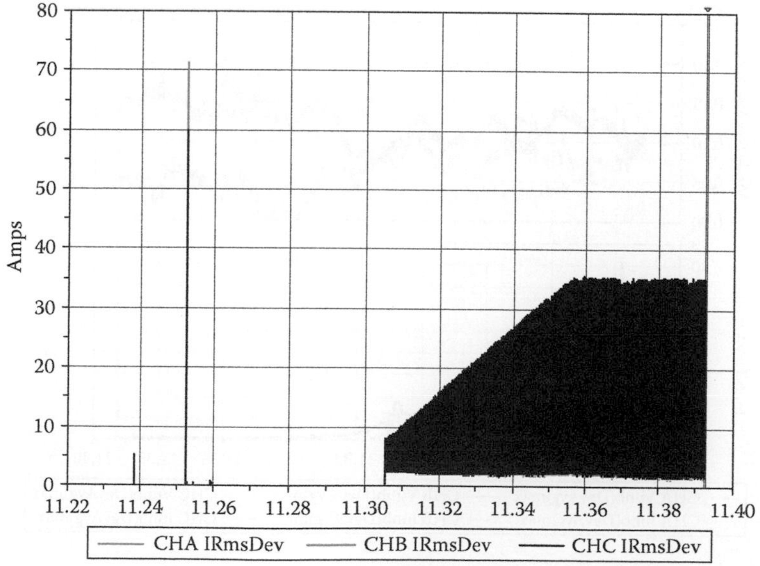


**FIGURE 6.12**
Approximate representation of the current unbalance.

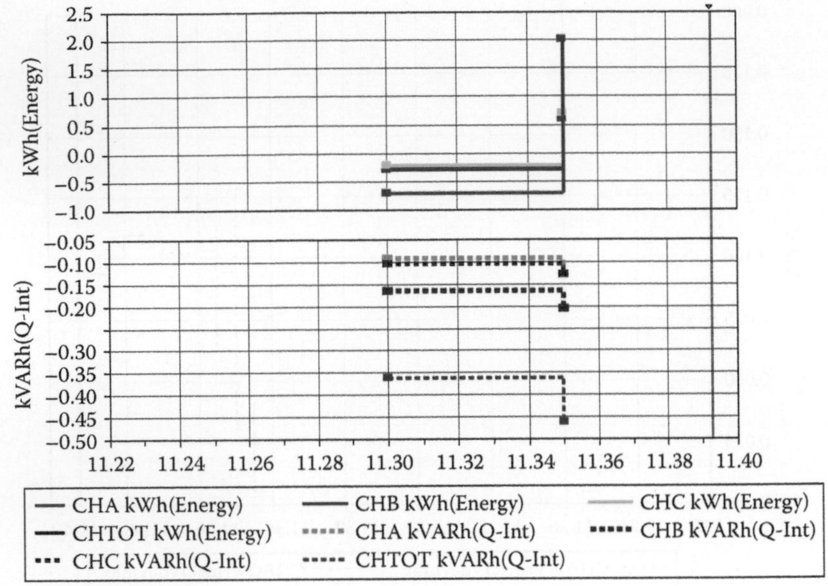


**FIGURE 6.13**
Approximate representation of the integrated active and reactive energy.

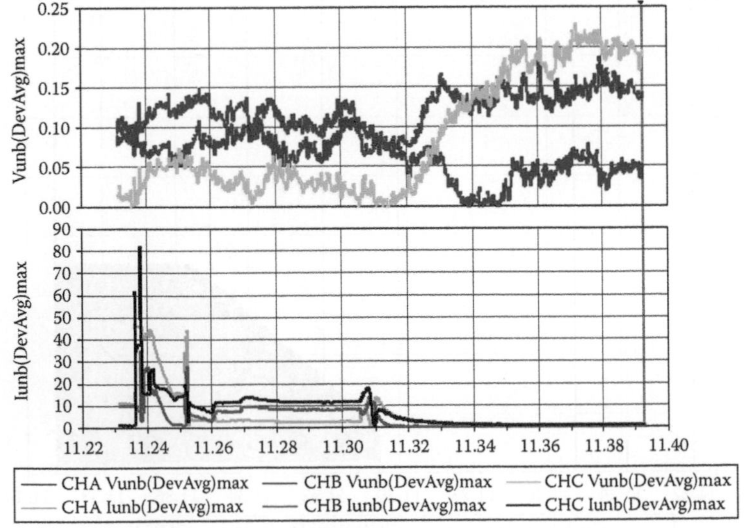


**FIGURE 6.15**
Approximate representation of the voltage and current unbalance (in%) according to the criterion of maximum deviation from the average.

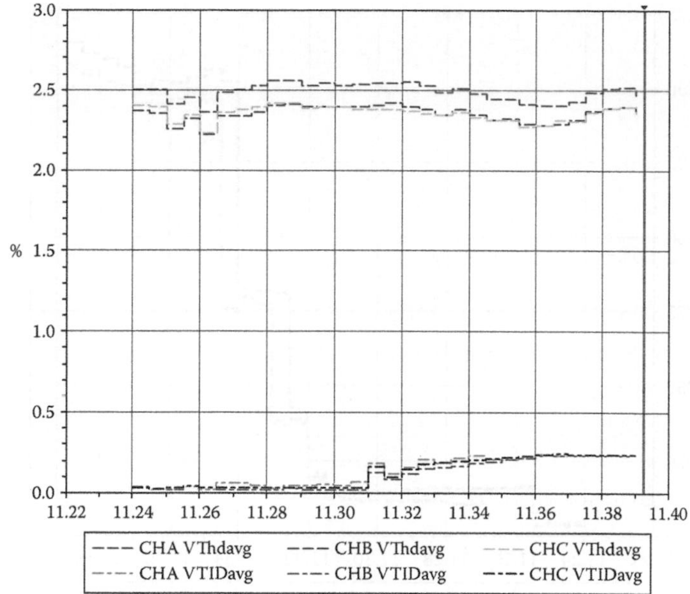


**FIGURE 6.17**
Approximate representation of the average voltage harmonic and average voltage interharmonic distortion (fundamental normalized).

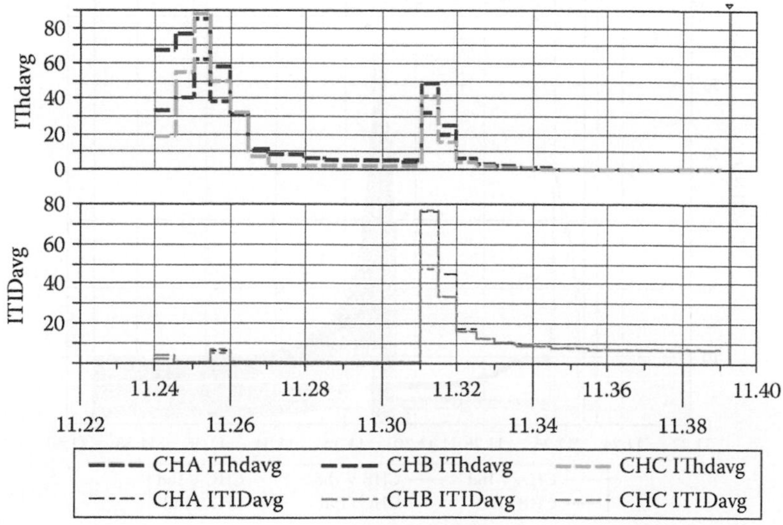


**FIGURE 6.18**
Approximate representation of the average current harmonic and average current interharmonic distortion (in%, fundamental normalized).

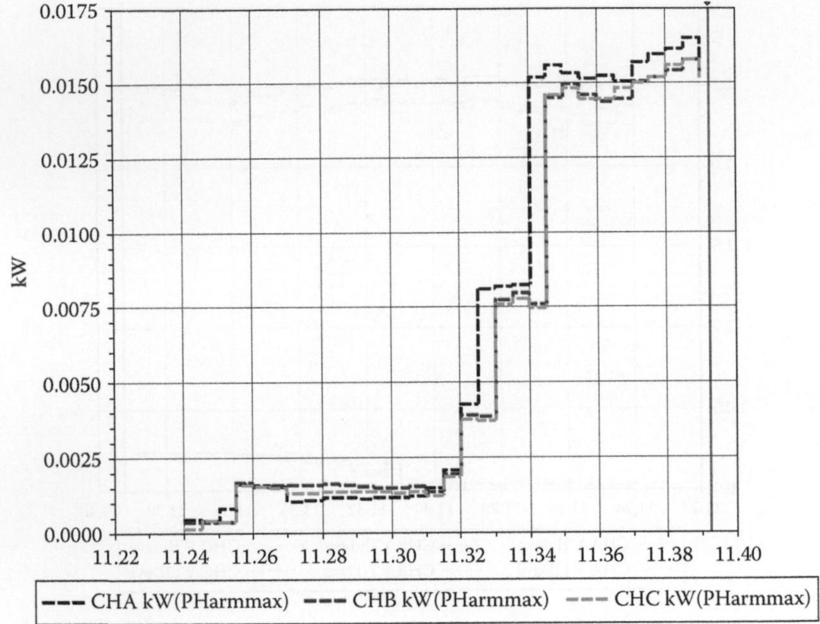


**FIGURE 6.19**
Approximate representation of the total phase harmonic power.

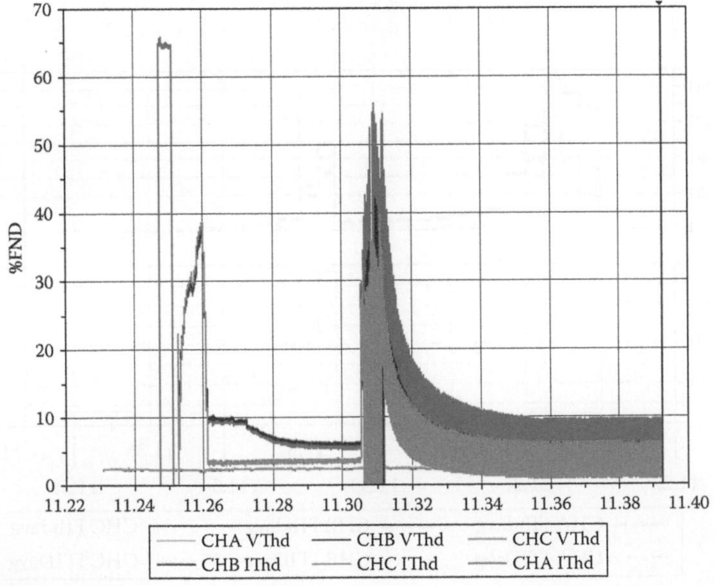


**FIGURE 6.20**
Approximate representation of the interdependency between the total harmonic content and the fundamental harmonic.

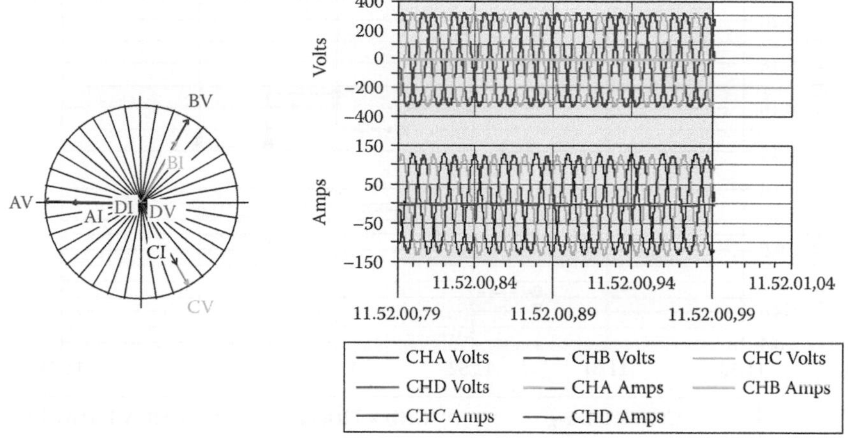


**FIGURE 6.34**
Approximate representation of the voltage and current waveforms prior to load variation.

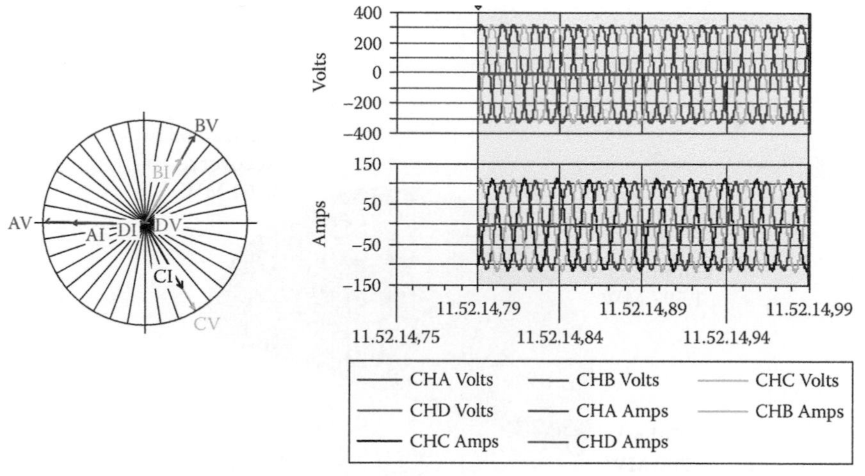


**FIGURE 6.35**
Approximate representation of the contribution and current waveforms after load variation.

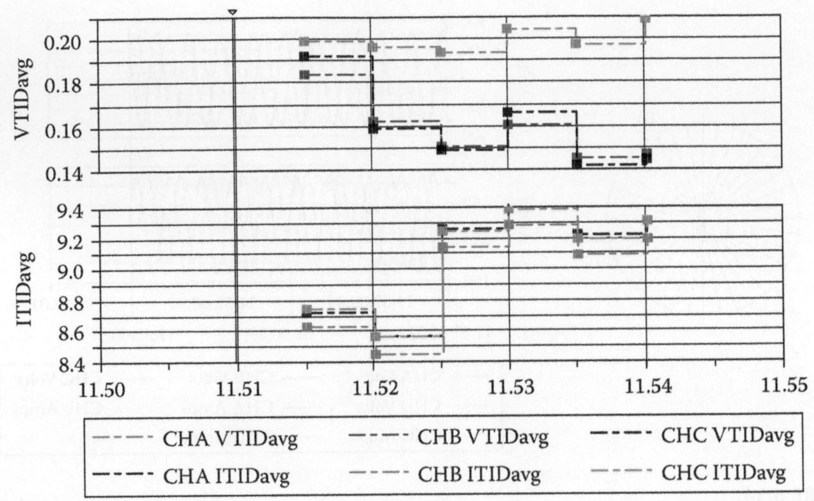


**FIGURE 6.42**
Approximate representation of the current average harmonic and current average interharmonic distortions (in%, fundamental normalized).

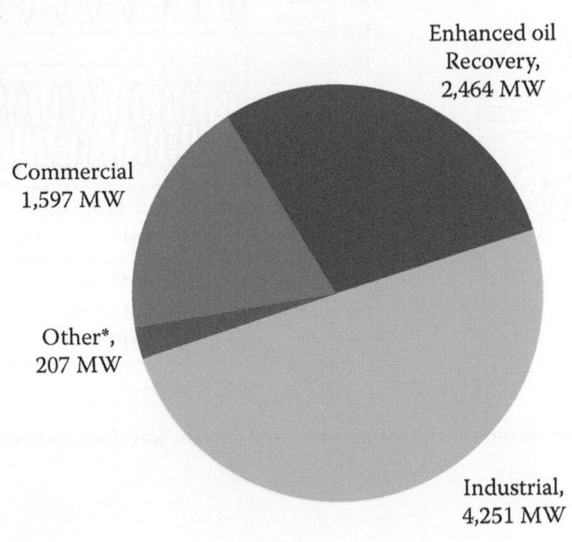


**FIGURE 8.1**
Existing CHP capacity in California according to the application class. (From Hedman, B., Darrow, K. Wong, E, and Hampson, A. ICF International, 2011. Combined Heat and Power: 2011–2030 Market Assessment. California Energy Commission. CEC-200-2012-002. With permission.)

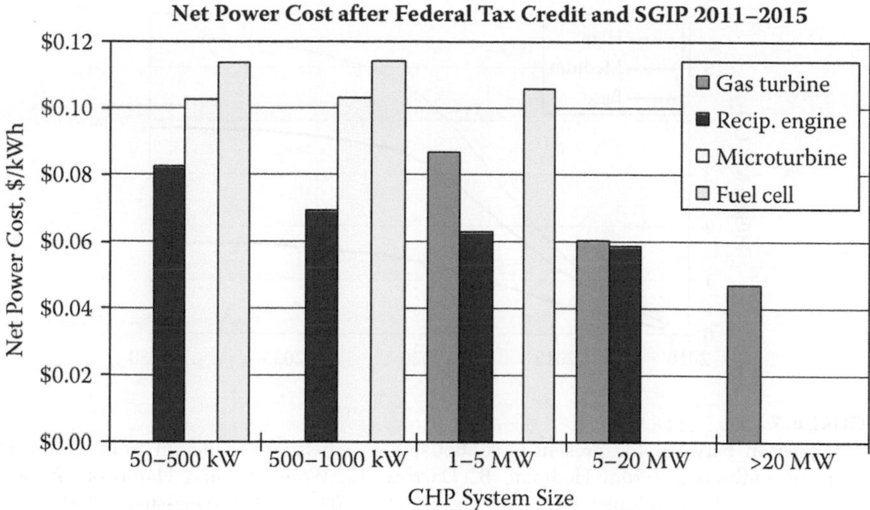


**FIGURE 8.2**
CHP net power costs as a function of technology and system size. (From Hedman, B., Darrow, K., Wong, E., and Hampson, A. ICF International, 2011. Combined Heat and Power: 2011–2030 Market Assessment. California Energy Commission. CEC-200-2012-002. With permission.)

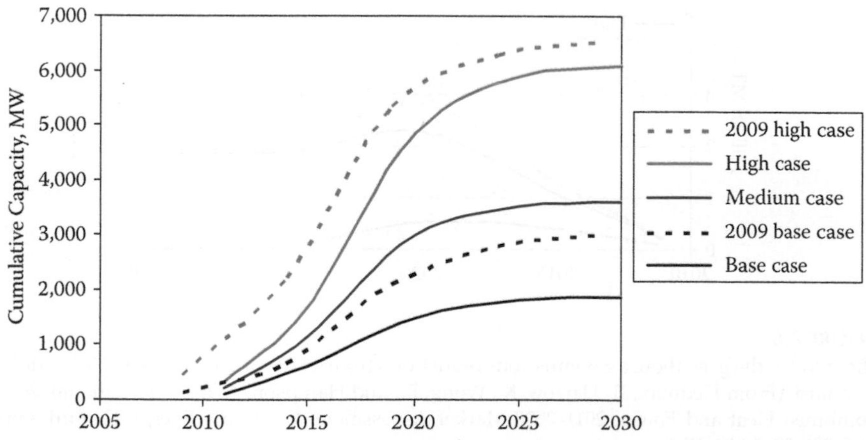


**FIGURE 8.3**
Cumulative market penetration as a function of scenario type. (From Hedman, B., Darrow, K., Wong, E., and Hampson, A. ICF International, 2011. Combined Heat and Power: 2011–2030 Market Assessment. California Energy Commission. CEC-200-2012-002. With permission.)

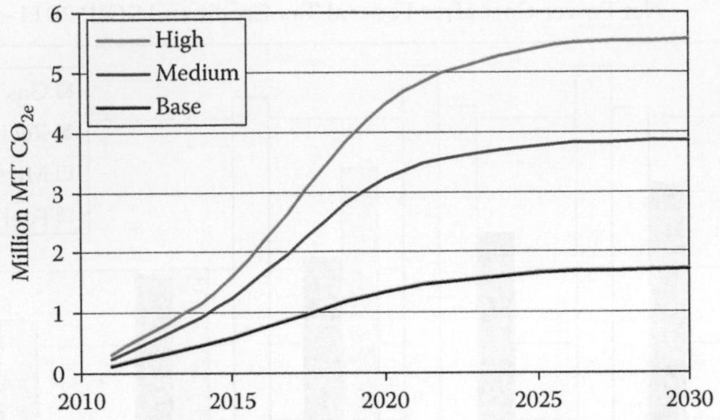


**FIGURE 8.5**
A comparison between the present Greenhouse Gas emissions reduction from CHP and the future emissions. (From Hedman, B., Darrow, K., Wong, E., and Hampson, A. ICF International, 2011. Combined Heat and Power: 2011–2030 Market Assessment. California Energy Commission. CEC-200-2012-002. With permission.)

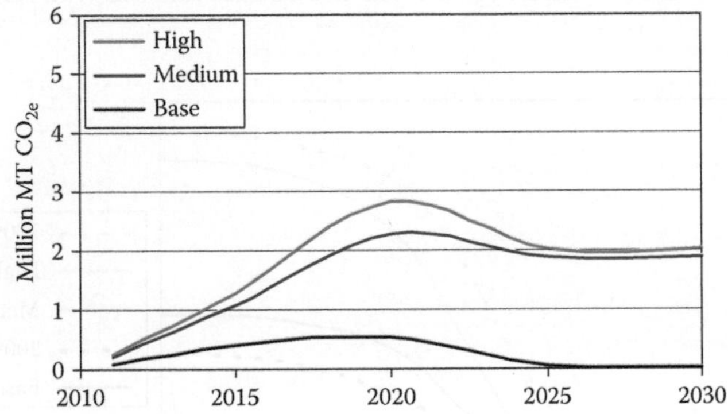


**FIGURE 8.6**
The trend of the greenhouse gas emissions reductions from CHP with 33% Renewable Portfolio Standard. (From Hedman, B., Darrow, K., Wong, E., and Hampson, A. ICF International, 2011. Combined Heat and Power: 2011–2030 Market Assessment. California Energy Commission. CEC-200-2012-002. With permission.)

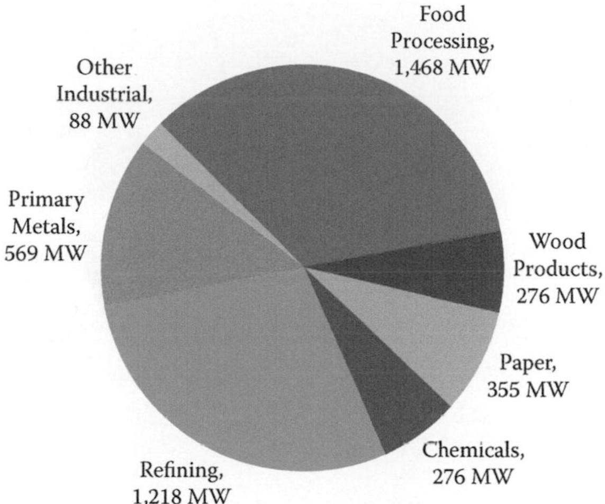


**FIGURE 8.7**
Current industrial CHP capacity in California. (From Hedman, B., Darrow, K., Wong, E., and Hampson, A. ICF International, 2011. Combined Heat and Power: 2011–2030 Market Assessment. California Energy Commission. CEC-200-2012-002. With permission.)

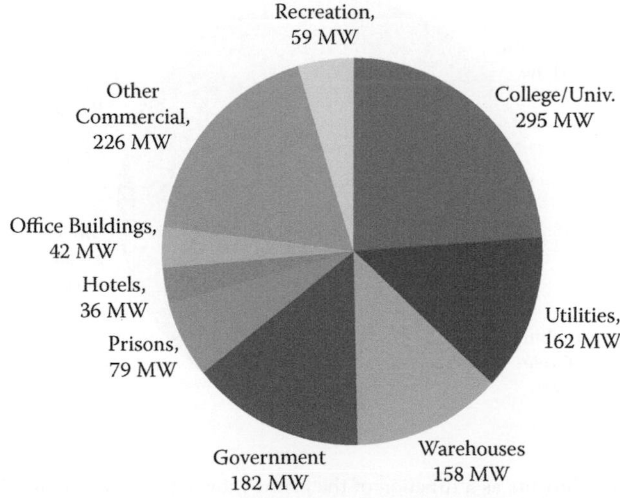


**FIGURE 8.8**
Current commercial/institutional CHP capacity in California. (From Hedman, B., Darrow, K., Wong, E., and Hampson, A. ICF International, 2011. Combined Heat and Power: 2011–2030 Market Assessment. California Energy Commission. CEC-200-2012-002. With permission.)

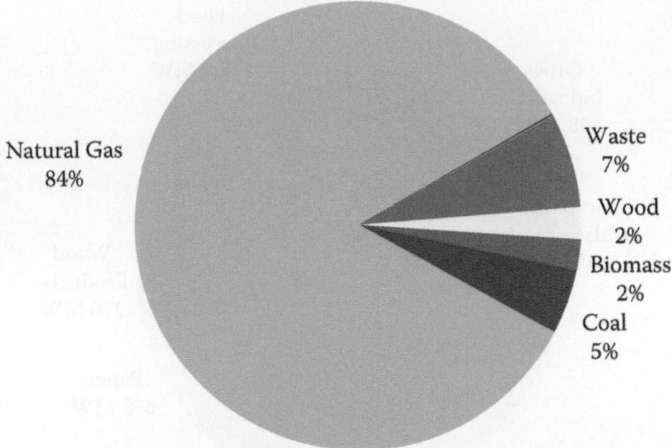


**FIGURE 8.11**
Existing CHP in California according to the fuel used. (From Hedman, B., Darrow, K., Wong, E., and Hampson, A. ICF International, 2011. Combined Heat and Power: 2011–2030 Market Assessment. California Energy Commission. CEC-200-2012-002. With permission.)

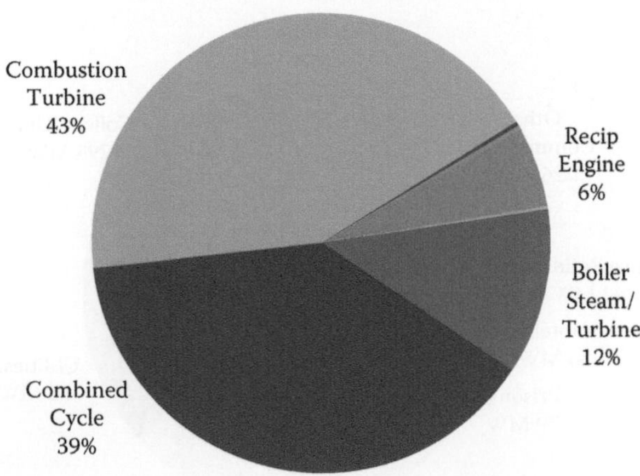


**FIGURE 8.12**
Existing CHP in California as a function of the prime mover. (From Hedman, B., Darrow, K., Wong, E., and Hampson, A. ICF International, 2011. Combined Heat and Power: 2011–2030 Market Assessment. California Energy Commission. CEC-200-2012-002. With permission.)

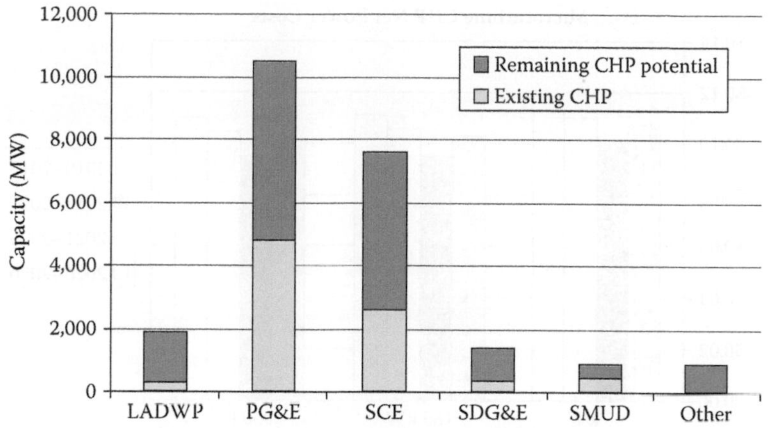


**FIGURE 8.13**
The present CHP potential and the total remaining CHP potential as a function of the utility territory. (From ICF International. With permission.)

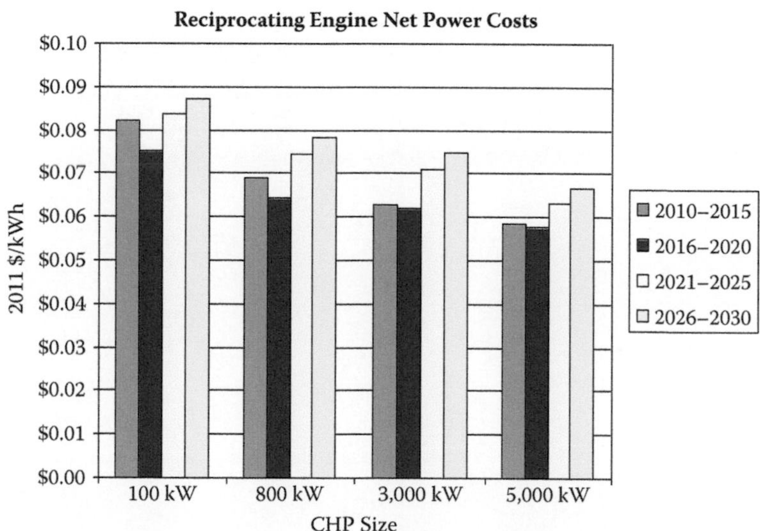


**FIGURE 8.16**
Net power costs for reciprocating engines. (From Hedman, B., Darrow, K., Wong, E., and Hampson, A. ICF International, 2011. Combined Heat and Power: 2011–2030 Market Assessment. California Energy Commission. CEC-200-2012-002. With permission.)

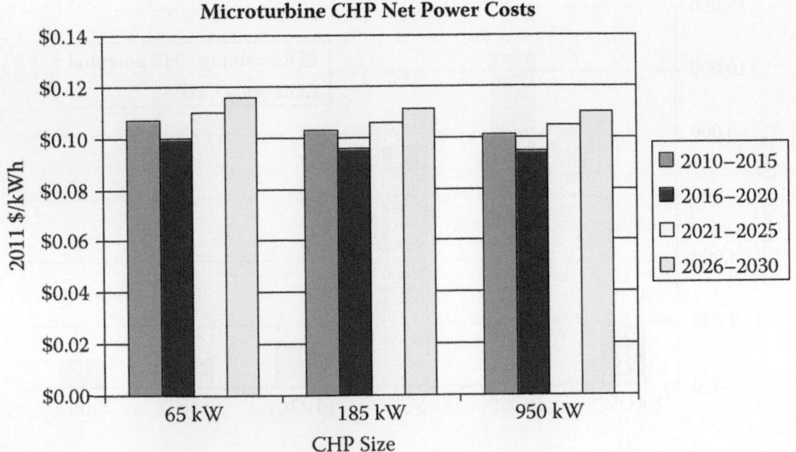


**FIGURE 8.17**
MT CHP net power costs. (From Hedman, B., Darrow, K., Wong, E., and Hampson, A. ICF International, 2011. Combined Heat and Power: 2011–2030 Market Assessment. California Energy Commission. CEC-200-2012-002. With permission.)

# 7

# *Case Studies*

One of the major advantages that characterizes microturbines (MTs) run on natural gas is flexibility. This means that they can be deployed in very different situations: either only for electrical energy production or in CHP (combined heat and power) and CCHP (combined cooling, heating, and power) applications. For a better understanding of the real behavior of a natural gas MT, several case studies from various countries have been selected.

## 7.1 Romania

### 7.1.1 Mediaş

The first unit was commissioned in Romania in 2011, was rated at 200 kW, and was installed at a hospital in the city of Mediaş. Its purpose was to generate both electricity and heat. It is estimated that the system will supply annually approximately 590 MWh of electrical energy and 2940 MWh of heat to the entire facility. Additionally, 750 MWh of electricity will be injected in the national grid on a yearly basis [CSO-- ].

The company that commissioned the MT affirms that the whole system will operate at an efficiency of a little over 90% and will permit the hospital to save around 10% of its annual energy costs.

## 7.2 Italy

### 7.2.1 Cavenago di Brianza

After 24 years of exploitation, the landfill at Cavenago di Brianza closed in 1994, and the whole area was transformed into a leisure destination. Far underground, the remaining decomposing garbage still generated large amounts of biogas containing methane.

Initially, the decision was made to employ a reciprocating engine, driven by the biogas produced by the landfill, to feed the headquarters of a company nearby. Eventually the methane content in the biogas diminished and thus could no longer be used for powering up the engine. Hence a more efficient, reliable, and environmentally friendly solution was searched for, and finally the decision was made to use five Capstone C65 MTs in a CHP application [CdB-- ]. Not only are they environmentally friendly, but they also have a good tolerance for siloxanes and hydrogen sulfide which are usually found in landfill biogas. The reciprocating engine did not present such characteristics and cannot be run on biogas with a methane content lower than 30%.

The five MTs have a rated capacity of 65 kW each, produce 325 kW (7800 kWh) of electricity every day which heats a greenhouse in the vicinity and powers up the leach installations of the landfill [CdB-- ]. During the summer, the heat energy is transformed to cooling for the air conditioning. The overall efficiency of the entire system is valued at a little over 80%.

The installation is made up of, in addition to the five MTs, a gas compression system that compresses the biogas that feeds the MTs, and a heat recovery system that minimizes the waste heat of each unit.

The initial reciprocating engine installed in 1970 was very noisy compared to the present MTs, and was a source of pollutants, while these ultra-low emission units reduce carbon dioxide by 3900 metric tons each year, and more importantly, can operate with smaller concentrations of methane.

### 7.2.2 Cossato Spolina

The wastewater treatment installation in Cossato Spolina represents another important example of waste reusing. Until spring 2009, the flame fed by the biogas resulting from this installation, burned above the entire facility. In time, the managers saw an opportunity to harness the energy of the biogas, and in spring 2009 a CR200 Capstone MT was installed. This provides all the electrical energy needed for the treatment plant as well as the heat necessary for the plant digesters to be operated at high efficiency [CSP-- ].

The MT produces 1.7 million kWh annually using 2600 $m^3$ of biogas from the plant, which initially was burned away as waste. An external heat exchanger coupled to the MT delivers another 2.3 million kWh of thermal energy to warm the digesters. Combined, the electrical energy generation together with the heat production will minimize carbon dioxide emissions by 1.8 tons a year.

As in the previous case, initially a reciprocating engine was used that ultimately turned out to be inefficient. The MT solution proved to have much lower maintenance costs and presented much better performance for hot water production.

The Cossato plant presently serves a cluster of several small cities totaling a population of 75,000.

### 7.2.3 St. Martin in Passeier

The two Capstone C65 MTs in St. Martin in Passeier, Italy, were commissioned in September 2010 and serve a hotel with a surface of 5000 $m^2$, totaling 150 rooms [SMP-- ].

Originally, the hotel depended 100% on utility power and on heat generated by boilers. A more economically viable and environmentally friendlier solution was searched for, which resulted in the commissioning of two Capstone C65 MTs.

The hotel implemented an aggressive waste recycling program, installing low-energy lighting built in accordance with the architecture [SMP-- ]. An annual $CO_2$ reduction of 800 tons was achieved. The entire system generates 125 kW of electricity each day, and the waste heat from the MTs is "recaptured" by special dedicated Capstone Heat Recovery Modules [SMP-- ] from which 224 kW of thermal power is produced and used to heat the swimming pools and wellness areas of the facility. The cogeneration system produces 1 million kWh of electricity on a yearly basis and 1.8 million kWh of thermal power, exceeding 80% energy efficiency.

The whole system will probably help the facility to cut 75,000 EUR of energy costs each year, reaching their return on investment in less than 3 years [SMP-- ]. It is estimated that 10,000 EUR of these 75,000 EUR will be the result of the low maintenance costs and reliability of the equipment.

Using the natural gas supplied by the public utility company, the two Capstone MTs cover approximately 50% of the facility's overall needs, replacing the initial boilers, which consumed more fuel and generated less thermal energy.

Despite the fact that the MTs are installed in a room next to the hotel, the noise is not loud. Fortunately, only 65 dB at 10 m are produced. Adding the fact that they are operated almost 8000 hours/year, MTs prove to be a very interesting and viable solution from all points of view.

## 7.3 Germany

### 7.3.1 Kupferzell

In Germany, at Kupferzell, the two Capstone CR 65 are fed with biogas resulting from biomass conversion [KAB-- ]. The conversion takes place through digesters that are filled with rotten vegetables and manure and then sealed. The result unleashes trillions of methane-producing bacteria that convert the worthless wastes into usable fuel [KAB-- ].

The German Biogas Association estimates that by 2020, 17% of the country's electricity will be produced by biogas. The most important advantage is that biogas can be very easily obtained, but a major problem is that many of the farmers are not able to find the best way to efficiently use this biogas.

One way is to adopt a similar solution to the one found at Kupferzell. Here, the Capstone MT® technology transforms organic waste to fertilizer which is used by the Kupferzell farms on a surface of more than 100 hectares. The first step was made in 2007 when a sludge drying system was purchased along with the first MT. At that time, two Capstone CR 65 MTs were commissioned and installed, converting pig and cow manure as well as food and crop residues into valuable fuel for generating electricity and heat [KAB-- ]. Two years later, the growing need for electricity forced farmers to upgrade to Capstone CR 200. In July 2009, the new CHP system was acquired.

The CR200 is presently operating only on renewable products, and generates 1500 MW of electricity and 2800 MW of heat on a yearly basis [KAB-- ]. This machine functions almost the entire year with an average availability of 97%. The MT feeds the electrical energy directly to the utility grid, while 100% of its waste heat is used to dry the sludge. Additionally, the low polluting emissions of the MT are diverted directly toward a dryer building for producing high-quality natural fertilizer resulting from the fermentation. Another advantage of this system is that an additional heat exchanger is not required because the exhaust is directly used for the drying, something that could not have been accomplished with a traditional genset motor. In the next stage of the process, the high-quality fertilizer is sold to local farmers and gardeners [KAB-- ].

The plant requires precise biogas flow analysis and measurements. The operators use a special dedicated measurement device and a SCADA (Supervisory Control and Data Acquisition) system, both developed by the German Capstone distributor. The SCADA system allows the monitoring, control, and supervision of the CHP application and facilitates other preventive maintenance actions as well.

Kupferzell receives a financial bonus from Erneubare Energien Gesetz (EEG), the German law agency, for renewable energies. Through EEG a certain electricity price for the supplied power is guaranteed, and a special separate technology bonus is granted for employing MTs. The plant meets and exceeds the standards set in the German air pollution control regulation "Technische Anleitung zur Reinhaltung der Luft" (TA Luft; "Technical Instructions on Air Quality Control") (Table 7.1) [KAB-- ].

### 7.3.2 St. Joseph Hospital

St. Joseph Hospital in Prüm was the first in Germany to install a Capstone C65 MT in a CHP application. After four years, in 2010, the overall impact proved to be outstanding [SJH-- ]. The entire CHP application achieves a total efficiency of 85% and saves the hospital 22,500 EUR annually in energy costs. Compared to the old boiler that served the facility for 60 years, the new Capstone MT helped to reduce energy use by 30%.

**TABLE 7.1**
Comparison between the TA Threshold and the Capstone CR200 Output

| Gas Type | Measurement Unit | TA Threshold | Capstone CR200 Output |
|---|---|---|---|
| Carbon monoxide (CO) | mg/m$^3$ | 100 | 69 |
| Nitrogen oxide ($NO_x$) | mg/m$^3$ | 150 | 47 |
| Formaldehyde ($CH_2O$) | mg/m$^3$ | 20 | 1 |

*Source:* Information on the Capstone microturbines at Kupferzell, Germany, http://www.microturbine.com/_docs/CS_CAP395_Kupferzell%20Biogas%20Plant_lowres.pdf (accessed April 1, 2013). With permission.

The CHP application delivers electrical and thermal energy to the hospital, which treats 5,000 patients each year, producing 60 kW of electrical power, 126 kW of heat, and 90°C hot water at its peak [SJH-- ]. In addition to 153 beds, surgery rooms, clinic spaces, and offices, the hospital offers a therapy pool and a physiotherapy practice. The new system operates more hours than the old boiler—approximately 7500 hours/year. The C65 run on natural gas generates between 40 and 60 kW of electricity, requiring only 8 hours of maintenance each year.

As in the previous case, there were some government incentives. Generally, the companies received about 0.1 EUR for every 30,000 kWh produced under the German Combined Heat and Power Law. Thus, the hospital decides to sell the electricity to the utility grid, it will receive 0.051 EUR for each kWh. According to the same law, the CHP units are exempt from the mineral oil tax, which saves the companies about 0.55 EUR/kWh [SJH-- ].

The installation of the CHP application at St. Joseph Hospital in Prüm was part of the government's goal to minimize greenhouse gas emissions by 23 million tons every year by December 2010. The hospital alone managed to reduce $CO_2$ emissions by more than 100 tons p.a. through the installation of the Capstone MT. By 2014, given the absolute success of this first unit, the hospital plans to transform the present application into a CCHP system through the installation of an additional absorption chiller.

### 7.3.3 The Q4C Oil Platform

The high reliability of the Capstone MTs stimulated an important oil and gas producer to develop the world's first North Sea platform designed specifically for MTs [Q4C-- ]. At present, the platform is fed with electrical and thermal energy by four C65 natural gas MTs which represent upgrades from the previously installed C60 units (installed in 2002).

The machines operate on wellhead gas, saving fuel transport costs to the platform. As in the previous case study, the low maintenance costs and requirements contribute as well to the economic efficiency. Compared to the classical reciprocating engines, which require at least four oil changes a year, the MTs need just an annual filter change and periodic routine inspections [Q4C-- ].

Additionally, the reciprocating engines have to be operated by maintenance crews that must be paid, transported to the platform on a helicopter, and the used oil has to be brought back to shore. These create substantial costs when compared to MTs. Here the important aspect of no lubrication need for the Capstone turbogenerators plays a vital role.

Regarding the operation, two of the four units function continuously, supplying 100–120 kW to the platform. The other two MTs provide backup power if needed [Q4C-- ]. The four MTs are cycled every 2 weeks when the backup units become prime movers, and vice versa. A Capstone APS (advanced power server) controls this cycling.

The success of the Q4C drew the attention of the other platform operators, and today many platforms in the North Sea are being fed with electricity and heat through MTs run on wellhead gas.

## 7.4 France

### 7.4.1 Cognac

The vinasse (a residual product) resulting from alcohol distillation necessary in the manufacture of cognac was until a few decades ago spread across the surrounding land [COG-- ], harming wildlife and polluting rivers in the vicinity.

In 1971, several distilleries joined forces to create a wastewater treatment plant whose role was to collect the vinasse, remove the polluting substances from it, and disperse it into nearby waterways. For the disposal of the 300 million liters of vinasse coming from 140 distilleries, the company installed four 5000 m$^3$ anaerobic digesters that break down waste matter. The water resulting from the process is separated from the sludge, treated, and released. During the breakdown process of the vinasse sludge, the bacteria in digesters generate large quantities of biogas, rich in methane [COG-- ]. The resultant methane was not flared directly into the atmosphere but was captured and used to fuel three boilers that produced steam, thus creating the electricity and heat needed to operate the facility.

As in the previous cases, a more efficient solution was sought and a Capstone C800 Power Package run on natural gas was commissioned in 2009. The role of this system is to provide electricity and heat for the company.

The C800 MT delivers 3000 MWh of electricity and 4000 MWh of thermal energy for the four digesters used in breaking down the vinasse. Another part of the generated thermal energy is directed to a greenhouse in the vicinity that grows flowers for the city of Cognac. The efficiency of the whole system exceeds 80%, and availability reaches 97%. C800 produces enough electricity, and the company is able to supply electrical energy to the main utility grid. According to the estimations, this will result in 400,000–500,000

EUR/year for the joint venture between the waste treatment company and the Capstone representative that installed the MT [COG-- ]. Initially, the system was considered so environmentally friendly that French and European officials offered a large subsidy.

C800 operates 9 months a year (from November to July), processing approximately 80 truckloads of vinasse during the peak season. To complete the process, the secondary products resulting from the digesters are mixed with green waste (leaves, branches, and flowers) from the Cognac region and placed directly on the soil to improve its quality.

### 7.4.2 La Ciotat

The situation at La Ciotat, France, was similar to the one at Cavenago di Brianza, Italy. The landfill operators from this town flared the methane resulting from waste decomposition of 20 years. As in many other places, the methane is no longer considered waste, and so in 2007 the decision was made to acquire 18 CR65 MTs which at present produce 1 MW of electrical energy each day, sufficient to power 1000 homes. As stated above, the MTs produce less pollutant emissions than the classical reciprocating engines. When compared to the same engines, the emission reduction in this case is equivalent to removing 700 cars from the road each day [Cio-- ]. Obtaining similar performances for these motors means adding other expensive equipment, and even in this case, the engines still generate pollutant emissions more than 10 times those of MTs. This is because the technology invented by Capstone does not use any lubricants, coolants, or other dangerous materials, since air bearings are employed. Theoretically, there is no need to filter the biogas feeding the MTs, but in this case it was decided to develop a filtering system that removes the siloxanes, hydrogen sulfide, and water from the landfill gas [Cio-- ]. As at Cavenago di Brianza, the landfill produces waste gas that contains less than 30% methane. This is also due to the fact that La Ciotat is old and located in a dry region. It is a well-known fact that aging landfills produce less methane, but despite this, MTs have proven to be very flexible, and studies show that they represent an interesting and viable solution for this type of location. Based on these positive results, some other 12 landfills in France and Belgium have benefitted from this equipment [Cio-- ].

## 7.5 The Netherlands

### 7.5.1 Rotterdam

The flexibility of MT technology was also proven in the Netherlands, where Capstone decided to create a unit with a rated power of 30 kW, which was installed on a vessel for inland shipping. This MT is fed with liquefied natural gas (LNG), being one of the first pieces of equipment of this kind installed on a ship operating on two types of fuel: diesel and natural gas [Rot-- ].

The vessel had to comply with the regulations of the Central Commission for the Navigation of the Rhine (CCNR), and thus its engines are now operating on a mixture of 80% natural gas and 20% diesel, attaining important $NO_x$, CO, $CO_2$, and $CH_4$ emission reductions [Rot-- ]. The ship relies on its two Capstone MTs not only for auxiliary power but also for further minimization of polluting emissions and fuel costs. This solution complies with the strict European emission regulations without any additional treatment, the result being an important reduction of the service requirements, operational costs, and maintenance [Rot-- ].

The exhaust of the MTs is reused through a heat exchanger to heat the water onboard. This hot water is then used to heat the LNG vaporizer, which delivers fuel to the two 30 kW units and the main propulsion engines. The hot water also produces thermal energy for the boilers (hot water domestic system), also being used for central heating [Rot-- ]. The air conditioning system is driven by an absorption chiller fed with heat from the MTs.

This CCHP application helps the vessel to improve overall fuel economy for auxiliary power, in this way also reducing the overall carbon footprint of the ship. In addition, the absence of any lubricant guarantees no pollution of the water.

The two units can also be run on natural gas, require minimal maintenance when compared to the classical diesel generators, and are very flexible as far as coupling with the existing equipment onboard. Other important advantages are very low noise levels and almost no vibration [Rot-- ].

## 7.6 Russia

### 7.6.1 St. Petersburg Region

In 2006, an onsite power system was installed at a sports center in the St. Petersburg region [STP-- ]. The 2.3 MW power plant consists of 30 Capstone C60 and 8 C65 MTs that are the single power source of the facility, which is located 54 km from St. Petersburg and the same distance from the nearest power line. The facility also includes a ski resort which is far from the utility grid, and the connection in this case would be very difficult and not viable from an economical point of view [STP-- ]. Another factor that contributed to the purchase of the Capstone MTs was fuel flexibility. Since the natural gas was very hard to obtain in this area, the chosen fuel was liquefied methane. Additionally, the 38 MTs can be operated on propane-butane. The needs of a ski resort are not uniform during the day, since at night the energy consumption is not so high. This aspect, together with the fact that these units comply with the most stringent regulations regarding pollutant emissions and noise, made these turbogenerators the best solution for this situation.

Officially, since September 2008, the MTs have covered all the electricity needs of the ski resort including the hotel, cottages, ski lifts, buildings, cafes, restaurants, and illumination of the slopes at night. The system generates thermal energy by capturing the exhaust gas resulting from the MTs. Thus, 4 MW of thermal energy are used for hot water generation and heating purposes.

### 7.6.2 Ukhta

The problem faced by a shopping center in Ukhta, Russia, was high utility bills. Through installation of MTs, utility costs were cut in half [Ukh-- ]. Another positive aspect was the fact that the Capstone system permitted a gradual increase of power output as the mall expanded and energy demand grew, thus reducing future capital investments.

Heat and electricity demand is covered in this case by ten C65 MTs and one C1000. These work around the clock, delivering power and heat to the 30,000 $m^2$ of the mall. The system was installed in three stages from 2008 to 2011, supplying 1700 kW of electrical energy and 2330 kW of thermal energy to the entire facility [Ukh-- ].

The efficiency of the whole system can attain 90% and meets strict regulations imposed by the government that require power efficiency of above 70% for all heat-generating installations.

The C1000 consists of five turbogenerators of 200 kW each, the internal redundancy of the system permitting maintenance operations to be executed on separate 200 kW units without shutting down the entire installation.

The return of investment for the first three C65 was expected to take place in just 2.5 years [Ukh-- ]. In the case of the C1000 and the other seven C65 units, the period is a little bit longer, approximately 5 years.

In this way, the energy independence of the mall, which represents an important facility in Ukhta, permits officials to concentrate on backing up the economy of the growing city and at the same time, of the region [Ukh-- ].

### 7.6.3 Mohsogollokh Village

Harsh weather conditions and an unreliable and expensive power grid obliged a cement factory in the Mohsogollokh Village to acquire two Capstone C1000 systems in a highly efficient CHP application. The extreme weather conditions usually reach –60°C in the winter, weakening the local central electricity grid such that frequent blackouts are the norm in the region [Moh-- ]. This situation has now been resolved. Today the Capstone CHP system provides thermal and electrical energy to the plant as well as to the 7000 residents of the region.

This 2 MW MT system has been in operation since January 2011, protecting the plant from the frequent blackouts of the past. The plant employs most of the residents in this region and is regarded as a community hero [Moh-- ].

The system also includes two UT-76 heat recovery units manufactured in Russia which produce 3.4 MW of thermal energy. The exhaust gases of the CHP application are directed to the common circuit of the boiler, which increases overall efficiency to almost 90%. These boilers provide heat not only to the cement plant but also to several buildings in the village, while the MTs deliver part of the generated electricity to these boilers, prolonging their life cycle and reducing maintenance costs [Moh-- ].

As in the previous case, the C1000 consists of five Capstone C200, and thus the entire system presents internal redundancy that permits the C200s to be placed separately out of service for maintenance purposes without shutting down the entire system. This feature provides round-the-clock operation, and above all, a much-needed heating supply to Mohsogollokh.

In this particular case, the central grid is used only as a backup for the MTs, thus saving money, as the cost for on-site generation of electricity is much lower than the local grid electricity rates. With an overall efficiency of 90%, reduced maintenance costs, and increased power reliability, the return on investment is expected to take place in just 3 years [Moh-- ].

### 7.6.4 Trolza ECObus–5250®

Another problem that Russia faces today is the growing number of polluting motor vehicles, and the need for environmentally clean and cost-efficient public transportation. The large electric trolleys, buses, and smaller streetcars are an important part of the country's public transportation system, but unfortunately they lack the capacity to meet Russia's accelerating transportation demands [Tro-- ].

The Trolza ECObus–5250 could be a viable solution. This bus combines the maneuverability of classic buses and the continuous running capability of an electric trolley or streetcar. Instead of relying on internal combustion engines like in the majority of the European countries, this bus makes use of batteries charged by an onboard Capstone C65 MT. The MT is fueled with natural gas stored in onboard reservoirs, and recharges batteries that have run low, allowing the bus to function without stopping for recharging [Tro-- ]. In this way, the distance the bus can travel without refueling is increased, and fuel consumption is reduced by some 40%. The lack of oil and cooling fluids makes maintenance more efficient and reduces operational costs.

The C65 designed for electrical hybrid vehicle applications can run on many different fuels, such as methane or diesel. Independently of fuel type, the MT complies with the strict Euro 4 emission standard, since the exhaust gases of the MT contain no more than 9 ppm of $NO_x$ and CO [Tro-- ]. This makes this type of bus a suitable method of transportation in leisure destinations as well as densely populated cities. Another advantage is that the polluting emissions are 12 times lower than those of the traditional diesel fueled buses and six to eight times lower than those generated by reciprocating engines [Tro-- ].

This means of transportation reduces maintenance, fuel consumption, and emissions and does not compromise passenger comfort. Through a heating fluid loop, thermal energy generated by the MT is captured and used efficiently to warm the passenger compartment without installing another autonomous heating system.

The noise level does not exceed 60 dB, which is comparable to that of a trolley. The compactness of the MTs has allowed engineers to increase passenger compartment space, and this bus is now capable of transporting 95 persons. This is why, at present, these eco buses are used in important cities and resorts in southern Russia, and in 2014 they will be employed for transportation at the Olympic Games in Sochi.

## 7.7 Bolivia

In Bolivia, where the government has recently nationalized gas fields and pipelines in order to increase state revenues, the decision was made to feed the gas compressors and pumping stations across the country using Capstone MTs. At the time of this writing, 11 units with a rated power of 65 kW each have been installed [Bol-- ].

More than 20 compressor stations push the natural gas through 6000 km of pipelines, and at many of these compressor stations the C65 machines assure the electrical energy supply. This is very important since Bolivia depends on the pipeline system to deliver the gas in its major cities and to generate income from exports to Brazil and Argentina [Bol-- ].

The company that adopted implementation of this solution had to install the MTs sometimes in very remote areas where the pumping stations are unmanned and require reliable and low-cost maintenance and round-the-clock operation.

The Brazilian pipeline and gas compressor system is powered by 40 MTs. This number is continuously growing in South America, and this type of unit gains more and more importance due to particularities of the landscape.

In Bolivia, the first two turbogenerators were commissioned in 2006 and replaced the classical generators, which were now obsolete from an electrical efficiency point of view. These two units were installed at Samaipata and Chilijchi. Since then, the rest of the MTs have been acquired and connected to the massive compressors that push the gas through the pipelines. The return on investment is expected to take place in 5 years [Bol-- ].

Another advantage is that these turbines use the raw gas flowing through the pipeline system, which makes them more efficient and increases the dependability of the delivery [Bol-- ]. The MTs present maintenance costs that are about 40% lower than those of a normal generator. At the time of this writing, the 11 turbogenerators have functioned more than 15,000 hours with impressive reliability. The end result is improved productivity and a shorter down time for maintenance and repairs.

## 7.8 Mexico

### 7.8.1 Gulf of Mexico

In the Campeche Bay on the Gulf of Mexico, there are 46 Capstone MTs that deliver electrical energy to 27 offshore oil rig platforms operated by a local company [GoM-- ]. Since 2002, this company has enlarged its fleet of Class 1, Division 2 Capstone MTs designed for hazardous locations. The decision was made, as in the other cases in this chapter, based on the fact that the Capstone turbogenerators meet the strict emissions regulations, maintain high reliability in dangerous environments, and have managed to uphold 2.5 millions of barrels per day in 2010 [GoM-- ]. These MTs, fed with wellhead gas, replaced the diesel generators that were not capable of withstanding the harsh, corrosive oceanic climate. Studies have shown that the C30s and C65s run safely in hazardous locations and require minimal space and limited maintenance.

## 7.9 United States

### 7.9.1 Simi Valley, California

The Ronald Reagan Presidential Library in Simi Valley, California, has a surface of approximately 9290 m$^2$ and gets 95% of its energy from 16 C60 Capstone MTs. These also deliver electrical energy to the Air Force One Pavilion, which is located nearby and houses the plane that flew seven U.S. presidents [Sim-- ].

The 16 MTs were commissioned in October 2005, and the system also includes four Carrier® absorption chillers. In addition to the production of 960 kW of electrical energy, the system delivers cooling and heating to the entire facility in a CCHP application. The exhaust gas from the turbogenerators is collected and diverted to the absorption chiller. This absorption chiller has a closed water loop and a chilling circuit that refrigerates the water that is subsequently used for the air-conditioning system [Sim-- ].

Since the system was installed in 2005, the MTs had an availability of 24/7 and were operated only on natural gas due to the lower impact it has on the environment than the other types of fuel.

### 7.9.2 Ravenna, Michigan

The CR30 that supplies the energy to a dairy in Ravenna, Michigan uses methane resulting from the manure produced by the 1000 cows of the farm as fuel [Rav-- ]. The heat resulting from the turbines is also reused for the separator building, having a surface of 65 $m^2$.

Initially, the manure is pumped through an external exchanger that heats the material to 38°C and then sends it to an approximately 14 m tall × 15 m wide digester [Rav-- ]. The digester from this dairy is anaerobic and uses no oxygen in the process; it features a continuous stir-tank reactor that mixes the preheated manure to break it down, thus creating methane, which is usually considered a waste material [Rav-- ]. Many anaerobic digesters in the world flare this waste gas, or even worse, disperse it directly into the atmosphere. It is well known that methane has a greenhouse gas effect on the atmosphere 21 times that of $CO_2$.

The digester from this dairy, manufactured in Austria, is the first of its kind in the United States. It has a few unique characteristics, including an inline $H_2S$ scrubber to remove the toxic hydrogen sulfide, pumps built into the digester foundation so that any sediment can be removed without emptying the tank, and separate units for easy operation and maintenance.

The Capstone CR30 produces 30 kW of continuous power, while the biogas from the digester also powers an 80 kW CHP reciprocating engine. The energy system at this dairy includes a 2.8 million BTU boiler. The heat exhaust from the turbogenerator is either diverted directly to the separator building or delivered to a heat exchanger to heat the hot glycol that provides heat throughout the entire facility. Before installing this system, the manure was stored onsite and eventually spread across the farm fields in West Michigan. The environmental agencies said that this was a problem due to the possibility of contamination of the water system by the manure.

An important advantage of the Capstone MTs is that they can be run on waste gases with low energy densities, and can accept $H_2S$ levels as high 70,000 ppm [Rav-- ]. The fact that the Capstone turbogenerators have only one moving part and require no lubricants or cooling fluids makes these units appropriate for this type of application.

### 7.9.3 Southern United States

Another important contribution of MTs was improvement of the electrical energy supply in hurricane winds of 190 km/h, which could cripple a key office or activity of the U.S. government [Gov-- ]. A classic example is a laboratory in the southern United States that analyzes various items to ensure national security.

The first problems arose in 2005 when Hurricane Katrina destroyed the previous location of this facility, and just 2 years later Hurricane Ike left the laboratory and the surrounding buildings without electricity for 1 month.

Due to these problems, in 2009 a new laboratory opened in the southern United States. The need for a more reliable power source was recognized immediately, and officials decided to acquire the UPSource from Capstone, an independent IT-grade power source that does not rely on the central grid, and also eliminates the need for large banks of DC storage batteries [Gov-- ].

The principal characteristic of this solution is its reliability. If even a single MT is lost, the whole system keeps almost eight-ninths of its reliability, which is much better than most other secure data sites [Gov-- ]. Another important advantage is that natural gas is more readily available during a powerful storm than the utility electricity.

The system includes six Capstone C65 units run on natural gas, two of these operating 24 hours a day, 365 days a year in a redundant configuration. The other units function at the laboratory to assure HVAC, and most importantly, power delivery during an extended outage. They operate synchronously and in parallel with the central electrical network so that the installation can detect an anomaly in grid stability. If this happens, the central grid is disconnected immediately and only a few critical consumers are still supplied with electricity [Gov-- ].

Five of the six MTs create the UPSource that supplies hot water to the laboratory and meets the heating requirements of the building. Each turbogenerator generates 251,000 BTU/hr of thermal energy, which is used for hot water production.

These ICHP (integrated combined heat and power) units have also removed the need for a secondary boiler system, so the system can function at efficiencies much greater than those of a traditional UPS (uninterrupted power source) or of a diesel backup generator [Gov-- ].

### 7.9.4 Manhattan, New York

Another location that has benefitted from MTs is a 35-story building located in Manhattan. The 12 C65 high-pressure turbogenerators installed on the 16th-floor setback roof of this building generate 780 kW of clean electricity, which represents approximately 35% of the building's total electricity needs [Man-- ]. The main idea was to reduce electrical energy consumption from the utility grid during peak hours and increase the power reliability of the entire building.

As in the other cases, the MTs deliver thermal energy that covers almost 80% of the location's total needs during the winter months. A heat recovery module is installed on top of each turbogenerator, capturing the exhaust heat that would otherwise have been dispersed into the atmosphere. Beginning in August 2006, this Capstone system has experienced almost 98.8% availability [Man-- ], which cannot be found in other DG systems.

Installing MTs is a way to increase the value of the building, as the owners will be able to attract higher-level tenants [Man-- ]. The installation operator must oversee that the total power costs of the building do not exceed the costs of remaining completely connected to the central electricity network.

Due to their improved reliability, lack of vibration, and lack of cooling and lubrication fluids, MTs represent a viable solution for powering all types of urban buildings [Man-- ].

### 7.9.5 Sheboygan, Wisconsin

The wastewater treatment plant in Sheboygan, Wisconsin, treats about 57 million liters of water a day, and consequently generates important quantities of biogas [She-- ].

The methane produced onsite in the anaerobic digesters initially had two purposes: to fuel an old 500 horsepower engine that was capable of pumping of 45,000 liters of water per minute, and to fuel the boilers that heat the digesters. Prior to 2006, 25% of this methane was considered waste and simply flared at the plant site.

The flared methane impacts 21 times greater on the environment than the CO, and so the process of finding a more reliable solution for reducing the polluting emissions as well as for generating electricity more efficiently has begun. Thus it was decided to develop a CHP application capable of generating both electrical and thermal energy for this plant. The most important advantage of a CHP application is that it is much more fuel efficient and environmentally friendly than the utility power and boiler heating. The reciprocating engine used to pump the water, on the other hand, needed 15 liters of oil each day to operate.

The system includes 10 C30 MTs, a gas cleaning installation that removes the moisture and siloxanes characteristic to landfills and waste treatment plants, and a gas compression system [She-- ].

According to the agreement between the company that operates the MT system and the city, the latter agreed to buy the electricity generated by the CHP application, install a heat recovery module, and provide the methane, while the former paid for all the components of the application. For every megawatt of clean energy the MTs produce, the plant receives one renewable energy credit that can later be sold [She-- ].

Today these units generate 300 kW of electrical power (2300 MW per year), as well as waste heat for the digesters and the plant building during the winter. The MTs recover 1 million BTU of heat per hour, enough to warm 60 homes in the city each year. These results are encouraging because the energy savings are significant, and the plant received more than $33,600 in revenues resulting from the energy generated by the MTs [She-- ].

### 7.9.6 Waynesburg, Pennsylvania

The company that operates 12,553 km of the natural gas pipeline system in the eastern United States made the decision in 2004 to decouple one of its transmission stations from the local utility grid [Way-- ]. The more reliable solution found at the time included three Capstone C65 MTs that provided all the electricity and heat needed.

If this location would have been fed with electricity from the utility grid, then the company should have provided power lines from a station 24 km away at a cost of $1.35 million and paid $0.116/kWh [Way-- ]. Through the installation of the MTs, the company saved more than $1 million, while at the same time controlling its own power source. The operator managed to easily pass the air quality tests since the turbogenerators present low $NO_x$ levels and practically no sulfur oxides. Therefore, in 2004 three Capstone C60 ICHP were commissioned, and in 2009 they were upgraded to C65 ICHP units.

The fleet currently consists of more than 46 units and generates more than 3 MW of electricity at 10 compressor sites across the eastern United States [Way-- ]. At the station that was equipped with MTs in 2004, these machines produce all the electricity and heat for the surrounding buildings, while the excess heat warms the natural gas, chilled during the decompression process [Way-- ]. Each MT contains a heat exchanger that warms the compressed natural gas to 30°C from an initial temperature of 2°C while the decompressed gas feeds the 7800 horsepower engines that drive the station compressors.

Another important advantage, as in the previous cases, is that these MTs, due to their excess heat, can play the role of a zero-emission, zero-fuel, 1 million BTU boiler [Way-- ]. In this way, the gas necessary to produce hot water becomes free. The efficiency of the entire system is estimated at approximately 85%, while the efficiency of the old reciprocating engine was about 30%, and the efficiency of the boiler had a value of 40%. The return on investment is expected to occur in 2014.

### 7.9.7 Carneys Point, New Jersey

Immediately after Hurricane Katrina swept the Gulf Coast in 2005, the local college in Carneys Point, New Jersey, was asked to continue its 15-year tradition of serving as the local Red Cross Disaster Relief Shelter [Car-- ].

According to the new agreement, the college must have a backup power system capable of supplying cooling, heating, and electricity to the county's emergency shelter, which has a surface of approximately 6039 m$^2$. The college officials decided to embark on the task and upgraded the power system to comply with the agreement. This happened through New Jersey's Public Utility SmartStart Incentive Program, and thus the college received a $130,000 grant to acquire the three Capstone C65 MTs (run on natural gas), a Capstone APS (Advanced Power Server), and a 100-ton dual burner absorption chiller [Car-- ].

Carneys Point is an interesting case due to the fact that before acquiring this MT-based CCHP application, it was heated through a geo-thermal system. Usually, MTs replace reciprocating engines. In this particular case, we have yet another demonstration of the fact that this emerging technology is very flexible.

Besides this geo-thermal system, two other not very efficient Freon™-based compressors and some natural gas boilers produced the necessary heat and hot water for the facility in the past.

The return on investment has been calculated to take place in approximately 10 years, though the officials expect that this interval will be shorter. Another advantage in this case is that the total energy savings are estimated at 30% while the overall efficiency of the CCHP application is valued at more than 80% [Car-- ].

The MTs are capable of working in islanding mode, thus assuring the electrical energy supply to the facility when the main grid goes down. Through these MTs, the cost for an additional generator (which otherwise remains idle most of the time) is saved.

The efficiency of the application is improved through the APS (advanced power server) as well. This is a stand-alone controller that supervises the load changes of the facility and automatically turns off the MT with the most operation hours when it is not needed; for instance, during the weekends or at the end of the day when there are not many students at school. The main objectives for this facility of reducing the costs for heating, cooling, and electricity have been in this way successfully attained.

# 8

# *Market Potential for Natural Gas Microturbines in California*[*]

After analyzing the essential aspects related to microturbine (MT) technology, the idea of a study that can quantify the evolution of the MT market at a global level would be tempting. In reality, such a report would not be possible since the CHP systems depend on the particular climatic conditions of a given area, and the global economy on a broad time horizon has many ups and downs. In practical terms, such a study would lack accuracy.

Nonetheless, ICF International prepared for the California Energy Commission in February 2012 a consultant report titled, "Combined Heat and Power: Policy Analysis and 2011–2030 Market Assesment," which quantifies the evolution of the CHP market in that time interval only for California [HKW12]. The remainder of this chapter has been entirely republished or adapted from this report.

**Disclaimer**

> This report was prepared as the result of work sponsored by the California Energy Commission. It does not necessarily represent the views of the Energy Commission, its employees, or the State of California. The Energy Commission, the State of California, its employees, contractors, and subcontractors make no warrant, express or implied, and assume no legal liability for the information in this report; nor does any party represent that the uses of this information will not infringe upon privately owned rights. This report has not been approved or disapproved by the California Energy Commission nor has the California Energy Commission passed upon the accuracy or adequacy of the information in this report.

This study analyzes the long-term market potential for combined heat and power (CHP) in California and to which extent the CHP can reduce potential greenhouse gas[†] (GHG) emissions. Market penetration estimates of CHP are presented for three market development scenarios—a base case reflecting the existing state policies and two additional cases (medium and high) that show

[*] This chapter has been entirely republished or adapted from [HKW12], with the kind permission of the California Energy Commission and of ICF International.

[†] There are a number of gases classified as "greenhouse gases" including carbon dioxide, methane, and nitrous oxide. This analysis only considers the impact on carbon dioxide considered to be the principal GHG produced from the deployment of CHP.

the market impacts of additional CHP policy actions and incentives. This study represents an update of a similar analysis that the team carried out in 2009.

## 8.1 Existing Combined Heat and Power Capacity in California

There are a certain number of databases about the current CHP projects in California that are maintained by the utilities, the California Energy Commission, California Public Utilities Commission (CPUC), and by the United States (U.S.) Energy Information Administration. At the same time, ICF International also supports a database of existing CHP for the U.S. Department of Energy (DOE). Of course the estimate of the total existing CHP applications in California differs among each of these sources. ICF reviewed the major data sources to develop an updated list of all existing CHP systems within the state. Based on this process, it is supposed that there are currently 8518 megawatts (MW) of active CHP in California at 1202 sites as shown in Figure 8.1.

### 8.1.1 Technical Potential for New Combined Heat and Power Capacity

The industrial, commercial, institutional, and multifamily residential markets have been analyzed in order to quantify the remaining technical

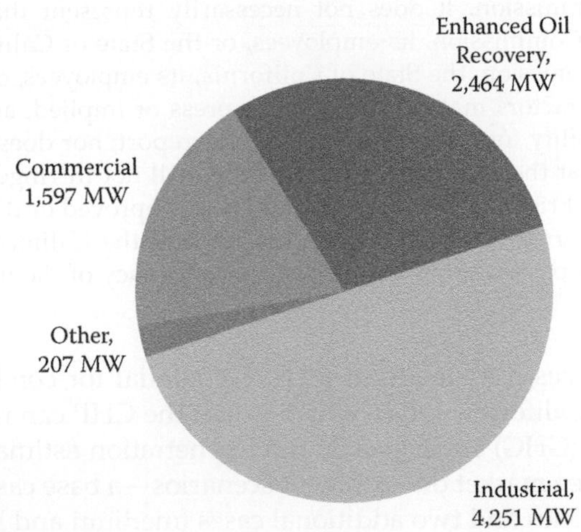


**FIGURE 8.1 (See color insert.)**
Existing CHP capacity in California according to the application class. (From Hedman, B., Darrow, K., Wong, E., and Hampson, A. ICF International, 2011. Combined Heat and Power: 2011–2030 Market Assessment. California Energy Commission. CEC-200-2012-002. With permission.)

**TABLE 8.1**
Technical CHP Potential in Existing and New Facilities as a Function of System Size and Market Segment

| Market Type/Size Category | 50–500 kW | 500–1000 kW | 1–5 MW | 5–20 MW | >20 MW | Total |
|---|---|---|---|---|---|---|
| *Remaining Technical Potential in Existing Facilities* | | | | | | |
| Industrial—on-site | 688 | 375 | 1,042 | 818 | 385 | 3,309 |
| Commercial, institutional, government, multi-family—on-site | 2,078 | 846 | 1,650 | 929 | 447 | 5,950 |
| Export | 0 | 0 | 286 | 901 | 3,847 | 5,034 |
| Total—existing facilities | 2,766 | 1,221 | 2,987 | 2,648 | 4,679 | 14,293 |
| *Technical Potential Related to New Facilities and Growth 2011–2030* | | | | | | |
| Industrial—on-site | 60 | 29 | 68 | 51 | 20 | 228 |
| Commercial, institutional, government, residential—on-site | 471 | 191 | 384 | 154 | 64 | 1,264 |
| Export | 0 | 0 | 9 | 40 | 131 | 180 |
| Total—new growth | 531 | 220 | 461 | 245 | 215 | 1,672 |
| Total | 3,297 | 1,441 | 3,439 | 2,893 | 4,894 | 15,965 |

*Source:* Hedman, B., Darrow, K., Wong, E., and Hampson, A. ICF International, 2011. Combined Heat and Power: 2011–2030 Market Assessment. California Energy Commission. CEC-200-2012-002. With permission.

potential for CHP. The technical potential is equal to the sum of the estimated new CHP capacity which could be built in applications that have the technical requirements (thermal loads, load factor, and size) necessary to support a potentially economic CHP project. The most important role is played by the thermal load. Most of the applications in the industrial sector having thermal to electric load ratios greater than one are sized according to the thermal load, and the excess power will be injected to the grid. When this ratio is less than one, all of the generated power on-site will be used. A summary of the technical market potential is shown in Table 8.1. As can be observed, there are 14,293 MW of remaining potential in existing facilities and an additional 1671 MW from expected business growth over the next 20 years. Of this total, 5212 MW represents the portion of the capacity destined for the export market. This capacity is concentrated in systems larger than 20 MW.

### 8.1.2 Combined Heat and Power Technology Cost and Performance

The cost and performance of the CHP technologies determine in great measure market response as well as economic competitiveness. The economical efficiency of the CHP applications is based on replacing the purchased electricity and boiler fuel with self-generated power and thermal energy. The savings in power and fuel costs have to be compared to the fuel, the added capital, and other operating and maintenance costs associated with operating a CHP installation. In this study, the cost and performance of primary CHP technologies have been evaluated as well as the technologies that are used in California including reciprocating internal combustion engines, fuel cells, gas turbines, and MTs. As shown here, only reciprocating engines and MTs have been investigated in detail. Twelve applications from 100 kilowatts to 40 megawatts were analyzed in terms of electric efficiency, thermal output, capital cost including emissions after-treatment costs, non-fuel operating, and maintenance costs. Figure 8.2 shows the estimated net power costs* for these systems using current energy prices. The study shows that in California the reciprocating engines tend to have most efficient technology in terms of costs, for sizes up to 5 MW. For more than 5 MW, gas turbines are the most economic technology. Emerging technologies such as fuel cells and MTs present higher net power costs but receive a certain market share due to benefits such as technical innovation, low emissions, and in the case of fuel cells, higher incentives.

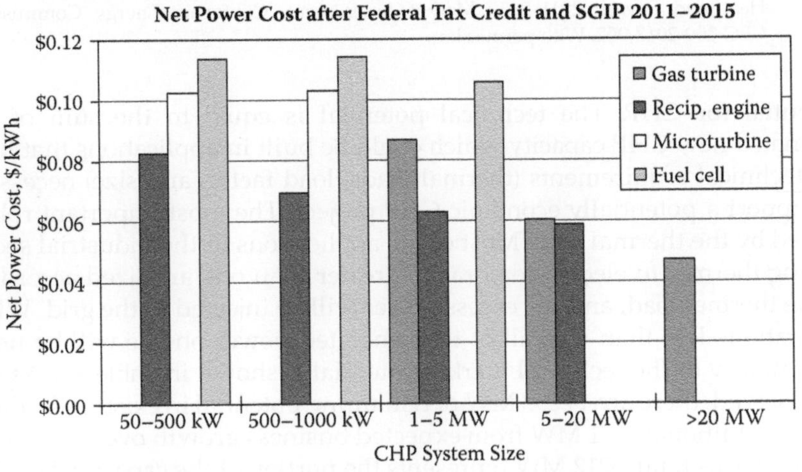


**FIGURE 8.2 (See color insert.)**
CHP net power costs as a function of technology and system size. (From Hedman, B., Darrow, K., Wong, E., and Hampson, A. ICF International, 2011. Combined Heat and Power: 2011–2030 Market Assessment. California Energy Commission. CEC-200-2012-002. With permission.)

* Net power costs represent the sum of the amortized capital costs (brought at the same level) at 10% return, the net increase in fuel costs after avoided boiler fuel and the operating and maintenance costs is subtracted—on a USD/kW basis. The value thus resulted is equal to the avoided cost of power that would provide a 10% rate of return.

### 8.1.3 Market Penetration Scenario Assumptions

This study analyzes market penetration of the new CHP installations over a 20-year time horizon (2011–2030). The base case describes the policies as they are expected to be implemented under current and emerging regulations:

- 33% Renewable Portfolio Standard—Most recently amended by Senate Bill 2 (Simitian, 2011)*, and CPUC Proceeding R.11-05-005—requires utilities to have 33% of their total generating capacity as renewable power by 2020.
- CHP Export Feed-in-Tariff—Assembly Bill 1613 (Blakeslee, Chapter 713, Statutes of 2007)—Fixes a price for the injection of excess power to a grid of an utility company from CHP facilities less than 20 MW.
- Qualifying Facility/Combined Heat and Power Settlement Agreement—CPUC Decision 10-12-035 December 21, 2010—resolved the disputes between utilities and qualifying facilities and established a new CHP procurement program through 2020. Though it is primarily focused on existing CHP applications, some terms and capacity limitations of the settlement influence the perspective for new CHP projects since the majority of CHP applications owners want to export power to the grid. The Short Run Avoided Cost Pricing mechanism adopted under the settlement agreement was used to represent the price paid for injecting power in the grid from projects larger than 20 megawatts.
- Cap and Trade—California Global Warming Solutions Act (Assembly Bill 32, Nunez, Chapter 488, Statutes of 2006)—provides a market trading program for carbon dioxide emissions which is designed to bring state emissions of greenhouse gases down to a 1990 level by 2020.
- Self-Generation Incentive Program—Senate Bill 412 (Kehoe, Chapter 182, Statutes of 2009)—revises and extends the program by inserting back the non-fuel cell CHP technologies and provides funding through December 31, 2015 as well.

The medium and high cases show that added CHP market penetration can be achieved with the following additional policy measures.

Medium Case

- Increase in market participation rates due to reduction in perceived market risk
- Large export markets (greater than 20 MW)

* For more information, please refer to http://www.leginfo.ca.gov/pub/11-12/bill/sen/sb_0001-0050/sbx1_2_bill_20110412_chaptered.html

- Pricing calculated according to the 2011 Market Price Referent which is 25–35% greater than the base case*,†
- Higher market response for paybacks less than 5 years
- Legislative extension of the Self-Generation Incentive Program (SGIP) beyond December 31, 2015, with programmed phase reduction in incentives until the payments become zero
- 10% reduction per year for fuel cells until dollar value of incentive equals the value of the other CHP technologies—then all technologies decline at the same 5% rate
- 5% reduction per year for all CHP technologies except fuel cells

High Case

- Standby power cost mitigation. In this case, the investor-owned utilities remove the non-bypassable charges which are currently applied to CHP systems and revise rates that require customers with CHP to simultaneously pay a standby reservation demand charge and additional demand charges for outages of the customer's generator. This change improves the savings from avoided electricity purchases by 1–2 cents/kWh.
- The Trade and Cap allowance costs for CHP fuel consumption are reimbursed after avoided boiler fuel is subtracted, eliminating the effective rise in natural gas fuel costs due to the Cap and Trade Program. It is assumed that Cap and Trade-related electric price increases are reimbursed on a 90% basis.
- Increased power generation from export projects by using combined cycle power generation technology for potential export projects greater than 50 MW. This change increases the large export technical potential from 3567 to 5401 megawatts, which is more than a 50% increase.
- 10% of the California investment tax credit is applied to CHP investments with no time limit or size restriction.
- $50 per kilowatt per year for transmission and distribution capacity and postponed payments for CHP systems with a capacity less than 20 megawatts.
- Capital Cost Reduction—an additional 10% reduction as far as the capital costs by 2030 concerns which is synonymous with the effect that higher market penetration will have on turnkey

* Based on Resolution E-4442, Public Utilities Commission of the State of California, December 1, 2011.

† For more information on the 2011 Market Price Referent Calculation Model please refer to: http://www.cpuc.ca.gov/NR/rdonlyres/B4F07AB3-0846-403B-ADDD-E6F495826113/0/Final2011MPR.xls.

design, technology improvements, and improved installation and interconnection practices.

- Increase in market participation rates in model analysis by an additional 2–7% compared to the medium case.

### 8.1.4 Market Penetration Scenario Results

Market penetration for new CHP capacity according to the three scenarios is shown in Figure 8.3 and Table 8.2. The 2011 20-year cumulative CHP market penetration ranges from 1888 MW in the base case to 6108 MW in the high case. The figure and table also compare the 2011 scenario forecast with the base and high cases from the 2009 CHP market assessment.

The study carried out in 2011 shows, in general, a lower cumulative market penetration than the study carried out in 2009. There are a number of contributing factors for this, such as the economic decline that has reduced the technical market potential, fewer existing businesses in California with CHP potential, and lower growth expectations for those markets over the next 20 years. The current CHP technology installation and capital costs used in the analysis have increased, on the other hand.

Another important aspect is that the export pricing for AB 1613 eligible projects had not been developed in 2009, and so the 2009 analysis was based on the renewable feed-in-tariff that includes a significant component related to avoidance of GHG emissions. The difference between gas and electric prices, often referred as the "spark spread," seems to be more favorable now than it was in 2009 due to a more favorable supply outlook for natural gas, but

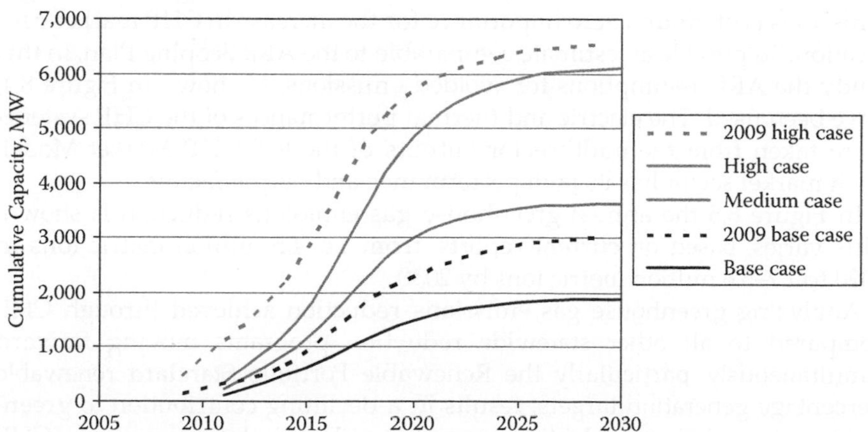


**FIGURE 8.3 (See color insert.)**
Cumulative market penetration as a function of scenario type. (From Hedman, B., Darrow, K., Wong, E., and Hampson, A. ICF International, 2011. Combined Heat and Power: 2011–2030 Market Assessment. California Energy Commission. CEC-200-2012-002. With permission.)

**TABLE 8.2**

Numerical Results According to Cumulative Market Penetration by Scenario

| | Cumulative New CHP Market Penetration, MW | | | | |
|---|---|---|---|---|---|
| **2011 Scenarios** | **2011** | **2015** | **2020** | **2025** | **2030** |
| Base case | 123 | 617 | 1499 | 1817 | 1888 |
| Medium case | 233 | 1165 | 3013 | 3533 | 3629 |
| High case | 340 | 1700 | 4865 | 5894 | 6108 |
| | **Cumulative New CHP Market Penetration, MW** | | | | |
| **2009 Scenarios** | **2009** | **2014** | **2019** | **2024** | **2029** |
| Base case | 136 | 680 | 2096 | 2816 | 2998 |
| High case (all-in) | 442 | 2209 | 5338 | 6306 | 6519 |

*Source:* Hedman, B., Darrow, K., Wong, E., and Hampson, A. ICF International, 2011. Combined Heat and Power: 2011–2030 Market Assessment. California Energy Commission. CEC-200-2012-002. With permission.

the benefits of lower gas costs are somewhat offset by GHG costs because of Cap and Trade. Moreover, Cap and Trade was not included in the 2009 study. In addition, the Self-Generation Incentive Program is more inclusive than in the 2009 report, but the stimulation of market penetration corresponding to the base case is limited by the current expiration date of 2016.

### 8.1.5 Greenhouse Gas Emissions Reduction from New Combined Heat and Power

The CHP contribution to the statewide reductions of greenhouse gas emissions is of ground zero importance for the increase in CHP market penetration. To provide an estimate comparable to the ARB Scoping Plan, in this study the ARB assumptions for avoided emissions, as shown in Figure 8.4, have been used. The electric and thermal performances of the CHP systems were taken from the multi-sector outputs of the ICF CHP Market Model. Each market sector has its own performance and output factors.

In Figure 8.5 the annual greenhouse gas emissions reduction is shown. This varies, based on current reports, from 1.4–4.5 million metric tons in 2020 to 1.7–5.6 million metric tons by 2030.

Analyzing greenhouse gas emissions reduction achieved through CHP compared to all other statewide reduction programs moving forward simultaneously, particularly the Renewable Portfolio Standard renewable percentage generation targets, results in a declining contribution to greenhouse gas emissions reductions over time. This is because on-site CHP reduces utility demand for electricity. Thus, this demand reduction in turn diminishes the amount of renewable energy capacity needed for utilities to meet their percentage targets. That is why, through the Renewable Portfolio Standard, the avoided utility emissions are valued at only 67% of avoided

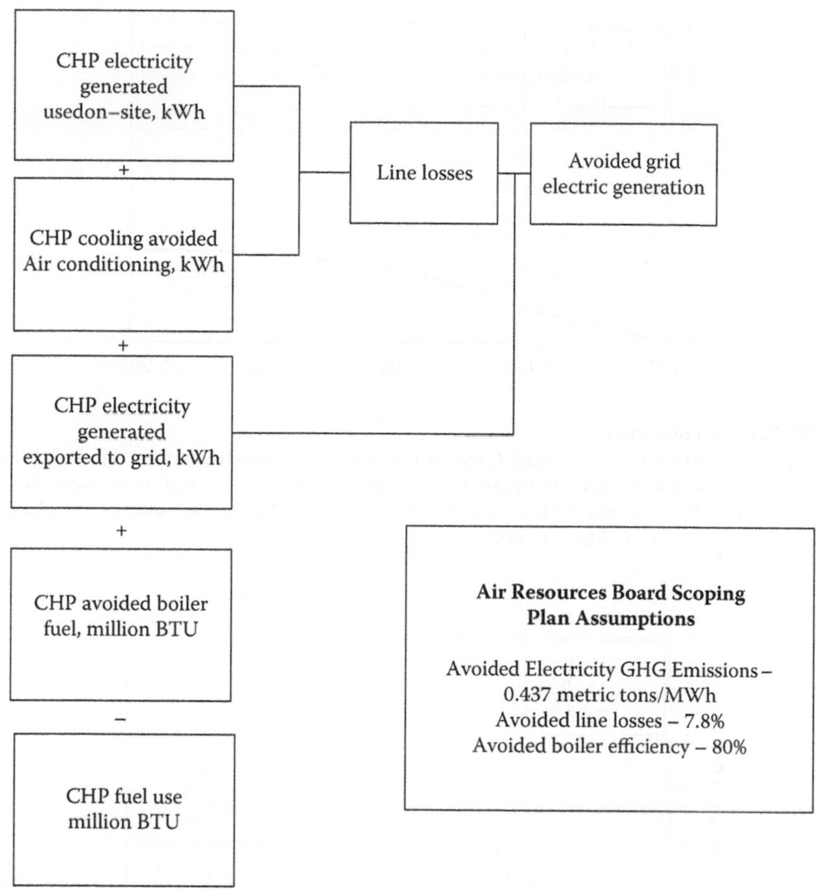


**FIGURE 8.4**
The estimation procedure for greenhouse gas emissions reduction resulting from CHP installations. (From Hedman, B., Darrow, K., Wong, E., and Hampson, A. ICF International, 2011. Combined Heat and Power: 2011–2030 Market Assessment. California Energy Commission. CEC-200-2012-002. With permission.)

emissions of the marginal fossil fuel electric system. Regarding the CHP that is exported, there is no reduction in GHG emission benefits because the emissions from the added CHP capacity are included in the estimation of utility greenhouse gas emissions or otherwise accounted for by the purchase of allowances by the export project.

Figure 8.6 shows the valuation of greenhouse gas emissions savings over time using the Renewable Portfolio Standard. The medium and high case reductions are lower than the base case because, as discussed, the export market penetration does not minimize the greenhouse gas emissions savings. The export market is much higher in the medium and high cases.

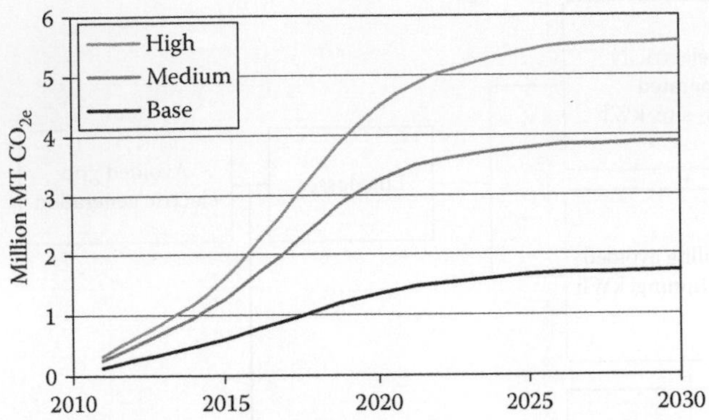


**FIGURE 8.5 (See color insert.)**
A comparison between the present Greenhouse Gas emissions reduction from CHP and the future emissions. (From Hedman, B., Darrow, K., Wong, E., and Hampson, A. ICF International, 2011. Combined Heat and Power: 2011–2030 Market Assessment. California Energy Commission. CEC-200-2012-002. With permission.)

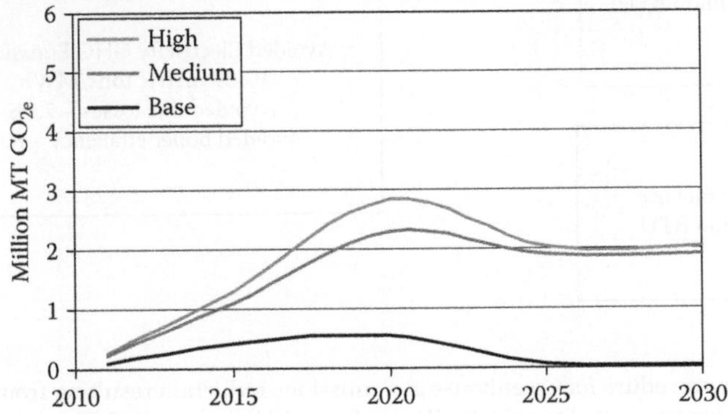


**FIGURE 8.6 (See color insert.)**
The trend of the greenhouse gas emissions reductions from CHP with 33% Renewable Portfolio Standard. (From Hedman, B., Darrow, K., Wong, E., and Hampson, A. ICF International, 2011. Combined Heat and Power: 2011–2030 Market Assessment. California Energy Commission. CEC-200-2012-002. With permission.)

### 8.1.6 Conclusions

The base case results show that given the current policy situation, the CHP will remain behind the ARB Scoping Plan market penetration target. Additional policy measures, mentioned in the medium and high cases, need to be undertaken in order to raise market penetration up to the Scoping Plan target.

In conclusion, the 2011 CHP market assessment shows a lower cumulative market penetration than the 2009 market study due to the following aspects:

- Lower export prices under AB 1613
- Higher costs for CHP installation
- Declined economic activity
- Gradual reduction or early ending of incentives under the Self-Generation Incentive Program
- Increases of the natural gas costs as a result of the cost of allowances under Cap and Trade

Another key factor is that the markets for large and small CHP systems present different needs and respond differently to various types of incentives. Table 8.3 points out the situation for large (greater than 20 megawatts) and small (less than 20 megawatts) CHP applications based on data gathered over 20 years. In addition, all markets benefit from investment tax credits. The small markets, above all, are negatively impacted by costs associated with Cap and Trade; large export markets can recover those costs by passing them on to the utility or by grouping them with the cost of power.

Table 8.3 also points out how important the stimulation of the export market is for achieving high levels of market penetration forecast through the medium and high cases. In the base case, the export market additions of new CHP are valued at only 213 MW. In the high case having higher pricing signals, market growth increases to 2457 MW. Prices having more or less the same value as marginal cost of power are needed for improved penetration of new large CHP export projects. Smaller eligible CHP projects having a lower capacity and higher costs make it difficult to compete even with the utility long-run marginal cost provided.

**TABLE 8.3**

Numerical Results for Cumulative Market Penetration According to the Market for Large and Small CHP Systems

| Scenario | Base | | Medium | | High | |
|---|---|---|---|---|---|---|
| Size | <20 MW | >20 MW | <20 MW | >20 MW | <20 MW | >20 MW |
| On-site | 1269 | 246 | 1519 | 263 | 2901 | 388 |
| Avoided air conditioning | 130 | 30 | 155 | 32 | 316 | 45 |
| Export | 91 | 122 | 93 | 1568 | 295 | 2162 |
| Total | 1489 | 399 | 1766 | 1863 | 3513 | 2595 |

*Source:* Hedman, B., Darrow, K., Wong, E., and Hampson, A. ICF International, 2011. Combined Heat and Power: 2011–2030 Market Assessment. California Energy Commission. CEC-200-2012-002. With permission.

The results of the export analysis in this study have been calculated, setting the price for export and letting the market model determine to what extent market penetration takes place. Given the Settlement Agreement and the Long Term Procurement Planning Process, the utility companies set the quantity of the CHP export desired while the price is determined by a bidding process. The targets of 3000 MW under the Settlement Agreement could be fulfilled by existing CHP systems, and after the 3000 MW target is accomplished, other new procurement targets will be determined in the Long Term Procurement Planning Process. That is why achieving the wanted levels of market penetration for new export CHP presented under the medium and high cases will be decisively influenced by the targets for the CHP capacity that are set.

The GHG emissions savings from CHP are smaller than the Air Resources Board target of 6.7 million metric tons per year of carbon dioxide, even in the high case where market penetration is better than the ARB estimate. The reasons for this difference are the result of the nature of the CHP markets. It has been assumed that all CHP market penetration has a high load factor system with full thermal utilization. In this study, thermal utilization rates for the small markets were equal to 80%. Instead, larger markets were assumed to have 90–100% thermal utilization. Moreover, those markets that use a certain percentage of the available waste heat to replace electric air conditioning have much lower emissions savings than those replacing strictly the boiler fuel. Markets having a low load factor save less due to their reduced annual hours of operation.

The other carbon reduction programs will contribute to the reduction of the marginal greenhouse gas savings over time when the Californian energy economy becomes less dependent on fossil fuels. The effectiveness of GHG emissions mitigation has to be based on cost analysis. CHP is less costly than some renewable energy sources providing equivalent emission reductions.

More important, CHP saves money for the facilities that adopt it. This is the main motivation that drives customer adoption. Until 2030, CHP would save customers $740 million per year in energy costs through the base case and $2.9 billion per year under the high case.

In 2006, California committed to reduce by 2020 its GHG emissions to the levels of 1990 by passing Assembly Bill 32, the *Global Warming Solutions Act of 2006* (Núñez, Chapter 488, Statutes of 2006). AB 32 was a decisive step for moving the Californian economy toward a sustainable, clean energy future. Being the lead agency responsible for implementing AB 32, the California Air Resources Board (ARB) prepared a comprehensive scoping plan that identified and presented an interesting approach to meeting this goal.* In this plan the ARB recognized CHP as an important

* For more details, please refer to Climate Change Scoping Plan: A Framework for Change, California Air Resources Board, December 2008.

component in regard to the GHG emissions reduction strategy. Another key point stressed by the ARB was the need for public policies to eliminate market and other barriers that are keeping CHP from reaching its full market potential.

## 8.2 2011 CHP Policy Landscape

The CHP policy framework changed a lot for both small and large CHP systems since 2009 when this study was first carried out:

- In 2010 and 2011, the CPUC released some decisions that have affected all Public Utility Regulatory Policies Act (PURPA) Qualifying Facilities (QFs) in California. The resulting CHP QF Settlement Agreement establishes a new Californian CHP Program, replacing the California PURPA program for CHP facilities having a capacity greater than 20 MW.
- Four relevant statutes were also codified:
  - Assembly Bill 2791 (Blakeslee, Chapter 253, Statutes of 2008) (AB 2791) added state, federal, and local government CHP facilities to the AB 1613 program
  - Senate Bill 412 (Kehoe, Statutes of 2009) (SB 412) improved the state's Self-Generation Incentive Program (SGIP)
  - Assembly Bill 1613 (Blakeslee, Chapter 713, Statutes of 2007) (AB 1613) allows the sale of excess power to a utility grid from CHP facilities of non-profit organizations
  - Assembly Bill 1150 (Perez, Chapter 310, Statutes of 2011) (AB 1150) extended the SGIP fund collection from December 31, 2011, to December 31, 2014, and maintained the administration of the fund through January 1, 2016
- Regulatory actions and related matters:
  - The California Air Resources Board (ARB) adopted its Cap and Trade program for the establishment, administration, and enforcement of a greenhouse gas allowance budget and provided a trading mechanism for compliance instruments (October 2011)
  - The standby exemption for CHP under 5 MW ended June 1, 2011
  - The CPUC in anticipation of the ARB Cap and Trade program released an Order Instituting Rulemaking (OIR, R11-03-012) on March 30, 2011, to approach among other things, the use of revenues resulted from the sale of GHG emissions allowances allocated to the electric utilities by the ARB

    - The CPUC set the stage for the Distribution System Interconnection Settlement (DSIS) process on August 19, 2011, to allow stakeholders a confidential forum to develop a revised Rule 21 which addresses, among other things, the interconnection problems associated with projects which will be exporting all or part of their excess power
- Non-bypassable surcharges on customer bills and their impact on CHP economics become of interest

Each of these events is summarized in the following text.

### 8.2.1 QF Settlement

On October 8, 2010, the *Qualifying Facility and Combined Heat and Power Program Settlement Agreement (QF Settlement)*[*] was filled and the CPUC quickly approved the Settlement (Decision 10-12-035, December 16, 2010). The role of QF Settlement, excepting the continuation of a PURPA Program for QFs 20 MW or less, set the framework for a State CHP Program as a replacement for the Federal PURPA Program. The approval of the Federal Energy Regulatory Commission (FERC) for the elimination of the must-take obligation for the non-PURPA Program was released on June 16, 2011.[†] More important is the following from the CPUC Decision:

> The Proposed Settlement is detailed. It would solve many QF problems and provide an organized transition from the existing QF program to a new QF/Combined Heat and Power (CHP) program. This new program is destined to preserve fuel efficiency, the resource diversity and the GHG emissions reductions, as well as other benefits and contributions of CHP. The Proposed Settlement is also destined for promoting new, lower GHG-emitting CHP facilities and encourage the operational changes and repowering, through utility pre-scheduling, or retirement of existing, higher GHG-emitting CHP facilities. Moreover, the Commission finds that the Proposed Settlement provides an improved allocation of the costs for the QF/CHP program to all customers in California who benefit from the CHP portfolio. The Proposed Settlement is comprehensive, but it does not resolve problems arising in numerous Commission proceedings implementing recent statutory requirements that correspond to QFs of 20 MW or lower, such as new CHP systems under Assembly Bill 1613 (codified as Pub. Util. Code sections 2840-2845), except to acknowledge that the megawatt (MW) and GHG reductions will be important for the investor-owned utilities' MW and GHG reduction targets.

[*] San Diego Gas and Electric Company, the Independent Energy Producers Association, Pacific Gas and Electric Company, Southern California Edison Company, the California Cogeneration Council, the Cogeneration Association of California, the Energy Producers and Users Coalition, the Division of Ratepayer Advocates of the California Public Utilities Commission, and The Utility Reform Network.

[†] Docket No. QM11-2-00.

The Settlement sets the stage for a new State CHP Program with a number of key elements and goals.* More importantly, it sets a target of 3000 MW of CHP, and a GHG emissions reduction target for the Electric Service Providers (ESPs), IOUs (investor owned utilities), and Consumer Choice Aggregators (CCAs) of 4.8 million metric tons (MMT).†

The Settlement enables a CHP facility, when nearing the expiration of its current Power Purchase Agreement (PPA), to consider a certain number of options: for instance, the CHP owner/operator could sell into the wholesale market, obtain a new PPA, shut down, or cease to export. The Settlement included several standard form contracts as well for existing and new CHP including: a CHP RFO (Request for offers) pro-forma PPA for new or existing facilities of 5 MW and larger that bid into a utility CHP-only RFO and win; a Transition PPA with avoided cost pricing for an existing QF with an expired or expiring PPA; an Optional CHP PPA for eligible As-Available Facilities; an Amendment for existing Legacy QF contracts; and a PURPA QF PPA for new and existing facilities of 20 MW or less.

According to the latest regulations, new and repowered facilities can benefit from a 12-year PPA, but will need to comply with some other additional criteria. There are also two PPAs for QFs under PURPA which qualify for an AB 1613 contract including one for a capacity of 5 MW and one for a capacity of 20 MW and less. Existing CHP resources that expand or repower which fulfill the criteria could benefit from different PPAs.‡

#### *8.2.1.1 PPAs for AB-1613 CHP 20 MW and below*

New or repowered CHP which comply with the technical requirements of AB 1613 can receive a Feed-in-Tariff (FIT) administered by the CPUC. The FIT is issued annually. The fixed charge paid is fixed-in per the PPA Term Start Date. The energy or volumetric charge varies each year and is adjusted according to the time of day of delivery, season of delivery, location bonus, gas price at utility's specified physical natural or gas delivery location. On the other hand, the price offered under the AB 1613 contracts is set in accordance with the costs of a new combined cycle gas turbine, and a location bonus is applied to eligible CHP systems located in local reliability areas.

#### *8.2.1.2 PPAs for AB 1613 CHP 5 MW and below (Simplified Contract)*

CHP installations of 5 MW, new or repowered, that comply with the technical and legal requirements of AB 1613 can benefit from a simplified

* For more information, see Section 1, CHP Program Settlement Agreement Term Sheet, dated October 8, 2010.

† This is based on the state-wide CARB Combined Heat and Power Recommended Reduction Measure of 6.7 MMT.

‡ These include AB 1613 PPA as well, less than 20 MW PURPA PPA, RFO PPA, and potentially others. Per Jennifer Kalafut, CPUC, email, October 20, 2011.

contract and from the CPUC administered FIT as well. The fixed charge paid is determined through the PPA term start date. On the other hand, the volumetric or energy charge varies each year and is adjusted as a function of time of day of delivery, season of delivery, a location bonus, and delivery location. The price offered through AB 1613 contracts is based on the costs of a new combined cycle gas turbine, and a location bonus is applied to eligible CHP systems located in local reliability areas.

#### *8.2.1.3 AB 1613 and AB 2791—Export of CHP*

The key role of AB 1613 and AB 2791 is to attract non-traditional customers to participate in CHP development, who otherwise would not have a budget for such projects: non-profits and federal institutions as well as state and local governments. The CPUC will establish in the future a pilot pay-as-you-save program for CHP systems with a capacity lower than 20 MW. This program will use on-bill financing, where the customer would have the capital and installation costs of a CHP system repaid by the difference between what would have been paid for electricity and the actual savings calculated for a period of up to 10 years. The pilot program has a 100 MW participation cap which is proportionately shared, based on the contribution to the state's peak demand, among the three investor-owned utilities. After finding no interest from affected customers, and complexities such as risks to ratepayers and application of federal and state lending laws in implementing the program,[*] the CPUC made the decision not to move ahead with such a program.

### 8.2.2 Self-Generation Incentive Program

During the interval in 2000–2001 when the electricity crisis caused outages throughout California, the legislature directed the CPUC to initiate certain load control and distributed generation (DG) program activities, including financial incentives to eligible customers.[†] The Self-Generation Incentive Program (SGIP) was formed to encourage the commercialization and development of new DG technologies.[‡] With the California Solar Initiative coming into being in 2006,[§] solar technology moved out of the SGIP into its own program. Today the SGIP is accepted as one of the largest funded and longest running DG incentive programs in the United States.

Right from the start of the program, the CHP was included as an eligible technology. Beginning in January 1, 2005, the combustion-based CHP using fossil fuel was required to comply with a stringent nitrogen oxide ($NO_x$) limit of 63.5 g/MWh, and on January 1, 2007, to meet the "ARB 2007" $NO_x$

[*] Decision 11-01-010 from January 13, 2011.

[†] For more information, please refer to Assembly Bill 970 (Alpert, Bowen, and Kelley, Chapter 329, Statutes of 2000) (AB 970).

[‡] For more information, please refer to Decision 01-03-073 from March 21, 2001.

[§] For more information, please refer to Senate Bill 1 (Murray, Chapter 132, Statutes of 2006) (SB 1).

limit of 31.75 g/MWh, regarded as the most stringent standard worldwide.[*] In 2006, the program was extended from January 1, 2008, to January 1, 2012 but had a limited eligibility of only wind and fuel cells.[†] In 2008, a California manufacturer became eligible for a 20% additional incentive.[‡]

In addition, in 2009 the CPUC was authorized to determine, in consultation with the ARB, what technologies should be included in the SGIP, based on GHG emissions reductions.[§] The SGIP was extended from January 1, 2012, to January 1, 2016. The long-awaited CPUC decision implementing the law was issued on September 8, 2011. However, due to the fund collection's rapid depletion in 2010 and the fact that funding was to end on December 31, 2011, the DG industry sponsored legislation that was enacted on September 22, 2011 that extended the fund collection of about $83 million per year for 3 years to December 31, 2014.[¶]

The CHP system manufacturers who put projects on hold since the passage of SB 412, effectively a 2-year period, were notified by the CPUC that they could begin submitting applications consistent with utility SGIP Handbook forms beginning on November 15, 2011. With natural gas forecasts that seem to be stable through 2030,[**] CHP systems are expected to be competitive with other eligible technologies.

There is an important difference between the previous and the latest SGIP, which is recognized as being budget weighted to renewables versus nonrenewable fuel technologies (75% compared to 25%). The most important characteristic of this SGIP is its hybrid performance-based incentive (PBI) with payments adjusted according to GHG compliance. In this case, 50% of the eligible incentive is paid up front. The remaining 50% is paid over a period of 5 years with the payment made based on performance which assumes a capacity factor of 25% for wind, 10% for advanced energy storage, and 80% for all other technologies. The payment characteristics for the GHG performance are:

- Half the payment in the years when the emission rate is between 398 kg/MWh and 417 kg/MWh
- No payment in any year in which the emission rate is greater than 417 kg/MWh
- A 5% exceedance band for GHG above 398 kg $CO_2$/MWh

---

[*] For more information, please refer to Assembly Bill 1685 (Leno, Chapter 894, Statutes of 2003) (AB 1685).

[†] For more information, please refer to Assembly Bill 2778 (Lieber, Chapter 617, Statutes of 2006) (AB 2778).

[‡] For more information, please refer to Assembly Bill 2267, 2008 (Fuentes, Chapter 537, Statutes of 2008) (AB 2267).

[§] For more information, please refer to Senate Bill 412 (Kehoe, Chapter 182, Statutes of 2009) (SB 412).

[¶] For more information, please refer to Assembly Bill 1150 (Perez, Chapter 310, Statutes of 2011) (AB 1150).

[**] For more information, please refer to ICF internal gas price forecasts.

Other important features include:

- A staged incentive for the first 3 MW, with decline beginning January 1, 2013, at 5% for conventional CHP
- An imposed minimum efficiency of 62% for higher heating value (HHV) of CHP systems
- Injection in the grid: 25% maximum of nameplate on an annual net basis
- Manufacturer's credit equals the unadjusted incentive (50 cents) × 1.2 for California manufacturers

The incentive levels by technology are presented in Table 8.4.

This update details the factors described above which affect the CHP economics for the market penetration study. Each of these incentives is paid according to the following rule: half at the beginning of the project and the other half as a PBI in equal installments over 5 years, depending on the system output. An example of PBI payment for a 3 MW CHP system is presented in Table 8.5.

**TABLE 8.4**

Incentive Categories and Values According to SGIP

| Technology Type | Incentive ($/W) |
|---|---|
| *Renewables and Waste Heat* | |
| Wind turbine | $1.25 |
| Bottoming-cycle CHP | $1.25 |
| Pressure reduction turbine | $1.25 |
| *Conventional CHP* | |
| Internal combustion engine—CHP | $0.50 |
| Microturbine—CHP | $0.50 |
| Gas turbine—CHP | $0.50 |
| *Emerging Technology* | |
| Advanced energy storage[1] | $2.00 |
| Biogas[2] | $2.00 |
| Fuel cell—CHP or electric only | $2.25 |
| CA Manufacturer's incentive | Unadjusted incentive × 1.2 |

*Source:* Hedman, B., Darrow, K., Wong, E., and Hampson, A. ICF International, 2011. Combined Heat and Power: 2011–2030 Market Assessment. California Energy Commission. CEC-200-2012-002. With permission.

[1] Stand-alone or connected with solar PV or any other eligible SGIP technology.

[2] Biogas incentive is an addition that may be used in conjunction with fuel cells or any conventional CHP technologies.

**TABLE 8.5**

Example of Performance-Based Incentive (PBI) Payment for a 3 MW CHP Using Natural Gas and Operating at an 80% Capacity Factor

| Year | Capacity (kW) | CF (%) | Hours/yr | kWh | Total kWh | PBI ($) | Total PBI ($) |
|---|---|---|---|---|---|---|---|
| 1 | 3000 | 80 | 8760 | 21,024,000 | 21,024,000 | 87,500 | 87,500 |
| 2 | 3000 | 80 | 8760 | 21,024,000 | 42,048,000 | 87,500 | 177,000 |
| 3 | 3000 | 80 | 8760 | 21,024,000 | 63,069,000 | 87,500 | 262,500 |
| 4 | 3000 | 80 | 8760 | 21,024,000 | 84,093,000 | 87,500 | 350,000 |
| 5 | 3000 | 80 | 8760 | 21,024,000 | 105,117,000 | 87,500 | 437,500 |

*Source:* Hedman, B., Darrow, K., Wong, E., and Hampson, A. ICF International, 2011. Combined Heat and Power: 2011–2030 Market Assessment. California Energy Commission. CEC-200-2012-002. With permission.

*Notes:* Calculation: $0.50/W incentive with staged incentive of 100% for first MW; 50% for the second MW and 25% for the third MW results in total of $875,000. Up-front payment of 50% of total, or $437,500. The remaining balance of $437,500 is paid over the last 5 years. [Note: If the CHP system is operated at an efficiency of more than 80% in a year, then it would receive the balance of $437,500 in a shorter period of time; but if it is operated at less than 80%, it only gets paid for actual kWh performance.] To determine the PBI payment for each kWh over 5 years, divide the total PBI by total kWh over 5 years = $0.004162/kWh.

### 8.2.3 Standby Rates

During the 1990s, it was expected that more commercial and industrial users would use the DG in the form of CHP and waste heat recovery. Thus, several DG groups formed to promote CHP. These were the Gas Research Institute, the California Alliance for Distributed Energy Resources, the Distributed Power Coalition of America (predecessor to the Gas Technology Institute), DG Forum, and the Electric Power Research Institute's (EPRI) Distributed Energy Resources. Unfortunately, at the end of the 1990s, the high natural gas prices and other tariffs provoked the decline of the CHP economics. The tariff design was disadvantageous to industry, utilities, and regulators equally. Much has been written about this issue.

> What does it cost the electric system to provide standby service for partial-requirements customers, and how should these costs be recovered? What are the benefits of DG to the system? How should standby rates be designed to reflect these benefits and encourage customers to maximize the value of DG for themselves and the system? The decisions made today will have long-term strategic consequences.*

The impact of standby rates on the CHP economics depends on their design (time-of-day [TOD] cost differences, seasonal variation, "demand ratchet," etc.)

* For more information, please refer to Johnston, Takahashi, Weston, and Murray, Rate Structures for Customers with Onsite Generation: Practice and Innovation, NREL/SR-560-39142. Executive Summary, page iii, December 2005.

and on the cost sharing between the fixed and volumetric charge components. Both fixed and volumetric charges represent a "cost of service," but there are many ways to calculate this and no one method seems to be correct.

Recovering the fixed costs in fixed charges makes lenders comfortable and contributes to the stabilization of the utility revenues but puts a heavy burden on small users and thus discourages investments in energy efficiency. Obtaining cost recovery through incremental usage encourages conservation but leaves the utility finances vulnerable to weather and other factors. The utility pricing must be designed in such a way that it should reflect the strategy of the times. The key element in designing this pricing must be represented by energy efficiency.*

California was one of the first states that exempted the CHP technology from the standby charges.† This exemption was intended to encourage customers to adopt the DG when the Californian electricity crisis in 2000–2001 provoked many outages. The initial exemption was valid only for CHP applications with a capacity of 5 MW and less which were installed before December 31, 2004. These CHP installations were exempt from the demand component of standby rates for a period of 10 years beginning May 2011. The exemption ended on June 1, 2011.

The CPUC developed its Standby Rate Design Policies for CHP systems with a capacity greater than 5 MW in 2001.‡ After this, the standby rate design was approached in each utility's General Rate Case. Unfortunately, whether the rates do meet the statutory requirements for customers using distributed energy resources is not clear. These requirements are:

(a) The tariffs required must assure that all net distribution costs for serving each customer class, considering also the actual costs and benefits of distributed energy resources, proportional to each customer cluster, as determined by the commission, are fully recovered only from that cluster. The commission shall require each electrical utility establishing those rates, to ensure that customers with similar load profiles within a customer cluster will, to the extent practicable, be subject to the same utility rates, regardless of their use of distributed energy resources. Clients having dedicated facilities are fully responsible for their obligations regarding payment for those facilities.

(b) The commission will prepare and submit to the Legislature, on or before June 1, 2002, a report describing its proposed methodology for the determination of the new rates and the process through which will establish those rates.

* For more information, please refer to Lazar, Jim, RAP, Challenges with Traditional Ratemaking, presentation. March 6, 2011. www.raponline.org/search/document-library/?keyword=Challenges+with+Traditional+Ratemaking&submit=Submit&publish_date_preset=&publish_date_start=&publish_date_end=&document_type_id=&sort=publish_date&order=desc.

† For more information, please refer to Senate Bill X1 28 (Sher, Chapter 12, Statutes of 2001) (SB 28).

‡ For more information, please refer to Decision 0107027.

(c) In designing the tariffs, the commission will consider the reliability of the onsite generation and the coincident peak load, as influenced and determined by the duration and frequency of the outages, so that customers with more reliable onsite generation and those who reduce the peak demand will pay a lower cost-based rate.*

And,

(d) The commission will maintain or adopt the standby rates or charges for combined heat and power applications which are based only on the assumptions which are supported by real data, and shall exclude any other assumptions capable of provoking outages or other reductions in electricity generation at combined heat and power systems which can occur simultaneously on multiple systems, or during periods of peak electrical system demand, or both.†

Recently, the utility company PG&E® negotiated an arrangement for designing the standby rate for non-residential customers, including standby rate design for the next three years.‡ SCE® and SDG&E® will probably revise their standby rate when they file their next General Rate Case application.

#### *8.2.3.1 Rule 21 Interconnection—AB 1613 Export Issues*

The CPUC jurisdictional Rule 21 regarding the interconnection process was originally intended to allow the interconnection at distribution level. In time, the state energy policy became more aggressive in mandating the adoption of distributed energy resources that need to be connected to the central grid using either the FERC jurisdiction Wholesale Access Distribution Tariff (WDAT) or the Rule 21 Tariff. This category of DG projects will employ programs like the AB 1613 feed in tariff for CHP§ or the CPUC Self-Generation Incentive Program (SGIP).

The CPUC took the decision of revising the Rule 21 Tariff to allow an increased number of interconnections and provide interconnection for projects that will entirely or partially be exporting power to the central network.

The CPUC also initiated on August 19, 2011 the Distribution System Interconnection Settlement (DSIS)¶ to provide the necessary means to evaluate the current CPUC jurisdictional interconnection rules and propose revisions to create a more transparent process. Accordingly, the DSIS team decided to finalize the technical framework for a revised Rule 21 tariff in

* For more information, please refer to PUC Code 353.13.(a) to (c).

† For more information, please refer to PUC Code 2841 (g).

‡ For more information, please refer to PG&E 2011 GRC, Phase 2.

§ For more information, please refer to http://www.cpuc.ca.gov/PUC/energy/CHP/feed-in+tariff.htm.

¶ Which was previously known as the Rule 21 Working Group.

the first quarter of 2012. At the same time, the CPUC will analyze the part of the DSIS agreement regarding Rulemaking R.11-09-011, whose role was to address the distribution system interconnection problems. It is believed that the DSIS settlement will provide an improved Rule 21 Tariff and that any issues between stakeholders that have not been resolved will be discussed in Rulemaking R.11-09-011.

#### *8.2.3.2 Departing Load Non-Bypassable Charges*

The departing load charges are approved and managed by the CPUC. They represent non-bypassable loads because the consumer who chooses to cover some of its load with self-generation systems cannot avoid the charges corresponding to this type of load.

An important aspect of this is that the non-bypassable charges have many components. Generally, these are based on the funding of public purpose programs for renewable resource technologies, self-generation, development and demonstration, energy efficiency, research, and low-income programs. Another category of these charges includes the nuclear decommissioning charges that were added by the Electric Industry Restructuring Law* or competition transition. Another charge was generated by the electricity crises of 2000 and 2001 that obliged the state of California to power procurement in order to cover the consumption not supplied by the utility companies of the state. Finally, the procurement costs resulting from the Department of Water Resources (DWR) were passed on to the customers of the investor-owned utilities as the DWR Bond Charge. Of course, these charges add costs to the CHP economics and negatively influence consumer decisions to adopt a CHP technology.

It is still under discussion whether the departing load charges should be reduced or even eliminated. It is clear that the charges affect CHP economics, and some officials argue that a reasonable reduction "would be lost in the rounding in remaining bundled customer rates."†

#### *8.2.3.3 AB 32 Carbon Cost Recovery—Cap and Trade Program*

> The three energy agencies from California have collaborated for the implementation of the Global Warming Solutions Act of 2006 (AB 32). Regarding the Cap and Trade, the 2008 Joint CPUC-CEC directives to ARB included the recommendation that ARB should treat the CHP operators comparable to retail providers for the electricity generated and

* For more information, please refer to Electric Industry Restructuring (Assembly Bill 1890, Brulte, Chapter 854, Statutes of 1996) (AB 1890).

† For more information, please refer to California Combined Heat & Power: Barriers to Entry and Public Policies for the Maintenance of Existing & the Development of New CHP. Slides 21–22. Michael Alcantar. Presentation at the Industrial Energy Consumers of America Meeting. June 21, 2011.

> produced on-site. On the other hand, the CHP operator should receive allowances on the same basis as retail providers and should be required to sell these allowances through a centralized auction organized by ARB or its agent and use the procedures according to AB 32.*

The ARB Cap and Trade carbon fee rules adopted October 2011 do not recognize CHP's avoided grid GHG emissions† and do not provide allowances to new CHP to offset GHG emissions. The rules require a carbon fee for the carbon emitted unless the facility is "trade exposed" (the cost of compliance will make the products of the facility more expensive than those of its competitors). In the case of the energy intensive trade exposed (EITE) facilities, the free allowances are allocated for a specified number of years. On the other hand, the CHP applications increase the on-site GHG emissions. As a consequence, the CHP owner must purchase additional allowances to cover these emissions, thus increasing his costs.

The CPUC proceeding regarding the utility cost and revenues associated with GHG emissions (CPUC R.11-03-012) has not yet been completed. The revised proposal on the appropriate use of allowance auction revenues to diminish the costs of AB 32 was filled on January 6, 2012 by the Joint Utilities. One interpretation of the proposal is that the allowances received by a customer as an IOU ratepayer (full requirements customer) cannot be kept if a customer chooses to install a CHP application (partial requirements customer). Furthermore, it is not clear if the customer who installs a CHP system can keep the allowance revenues corresponding to the remaining load fed by the utility company.

### 8.2.4 Continued Production from Existing QF/CHP

An anticipated CPUC approval of the CHP Program Settlement, awarded in July, would have led to the first utility solicitations (RFOs) in October. The final and non-appealable CPUC approval was not achieved until November 23, 2011 (which is referred to as the Settlement Effective Date). According to the terms of the Settlement, PG&E‡ and SCE§ launched their CHP RFOs on December 7 and 15, 2011, respectively. According to the terms of the settlement, each IOU will hold three CHP-only RFOs before the end of the Initial Program Period (November 22, 2015).¶

* For more information, please refer to D0810037, Order #22. Also see Findings of Fact 57, 58, and 59.

† The Cap and Trade regulations were in fact adopted at the CARB's October 20, 2011, Board Meeting.

‡ PG&E: December 7, 2011. For more information, please refer to http://www.pge.com/b2b/energysupply/wholesaleelectricsuppliersolicitation/CHP/CHP.shtml.

§ SCE: December 15, 2011. For more information, please refer to http://www.sce.com/EnergyProcurement/renewables/chp.htm.

¶ For more information, please refer to Section 5.1.4, QF CHP Program Settlement Agreement Term Sheet, page 27.

It has been anticipated that the industry has some experience with solicitations as well as with contract options. This was not the case, and many of the stakeholders refused to speak publicly during the development of contract offers. The surveys nonetheless revealed some perspectives, which are discussed next.

Plant Closures, Expansions, and Repowering

- The inefficient units are expected to repower, shut down, or convert to a Utility Prescheduled Facility (UPF).*
- The RFO prices are determined through the bidding process. Those facilities that remain on SRAC (short run average cost) are treated according to the Settlement SRAC which replaced the CPUC-adopted SRAC formula on January 1, 2012.†
- A CHP facility that is currently selling to an IOU under a Legacy PPA or an extension can sign a Transition PPA with the same IOU when the PPA expires during the Transition Period.
- Older and inefficient CHPs will be transformed into dispatchable units for economic reasons.

Request for Offers

- It is thought that the projects operating now will continue to operate. When the legacy contracts approach expiration, these QFs are expected to search for a new PPA.

The MW Target

- At this time, there is no preconception of how the target will be achieved. All the contract options available in the QF Settlement are expected to be used.
- The settlement does not specifically deny the advantage of seeking a QF contract, and there is consensus between the parties that if there are facilities listed in the IOUs' July 2010 semi-annual reports, then these contracts can be considered for the IOU's MW target.‡

Terms and Conditions

- It is not believed that the Settlement terms and conditions will be an issue for existing QFs because they were heavily negotiated in the QF Settlement.
- New facilities will likely have extended negotiations versus an existing facility due to the fact that there are many unknowns

* Utility Prescheduled Facility is in fact defined in the Settlement as an Existing CHP Facility that has changed operations to convert to a utility-controlled, scheduled, dispatchable generation facility, including but not limited to an Exempt Wholesale Generator (EWG).

† For more information, please refer to Section 10, CHP Program Settlement Agreement Term Sheet, dated October 8, 2010.

‡ For more information, please refer to Section 5.2.3, CHP Program Settlement Agreement Term Sheet, dated October 8, 2010.

regarding terms and conditions that would apply to a new plant and its intended operation.

- The dispatchable requirement is problematic for facilities that operate on fixed schedules or have to cover constant loads.

California Independent System Operator (CAISO) Interconnection Process

- The costs for the new California Independent System operator (CAISO) metering and software are not considered expensive. Being a participant in a cluster study instead, could take time and be costly.
- The CAISO review process is long, and this will impact the beginning of the operations.

Key Drivers Affecting CHP Market—Policy, Environmental, Economic, Technical, and Terms and Conditions

- Economic aspects as well as the GHG policy drivers depend decisively on whether the CHP facility is owned by the industrial host, or a third party.
- Some of the third-party owners of CHP applications have steam and/or retail electricity contracts with their hosts that predate the passage of AB32. Another important aspect is that many of these legacy contracts do not include provisions for GHG cost recovery, and the host customer has no incentive to renegotiate the contract.
- The Settlement only goes through 2020; a long-term plan to 2050 is needed.
- According to present assumptions, future industrial growth is either flat or negative. For the industrial sector, the market potential analysis for new CHP needs to make sense and be consistent with this growth rate.
- GHG reductions from CHP can vary a great deal depending on aspects such as CHP technology and whether all power is consumed on site or if a portion is sold to the grid. Given this, a MW target is not always appropriate if the goal is represented by GHG reductions.

## 8.3 Existing Combined Heat and Power Capacity Update

This study estimates that there are 8518 MW of operating CHP in California at 1202 sites. The existing CHP is defined in this assessment to contribute in both the characterization of the technical market potential for new CHP deployment and the evaluation of the barriers to continuation of existing CHP contracts under the QF settlement agreement. Data from several Californian specific sources were compared to ICF's CHP Installation Database.

The ICF's CHP Installation Database includes data on the CHP applications throughout the entire country in all size ranges. The database is constructed using a variety of sources such as the Department of Energy (DOE), the Energy Information Administration (EIA) electricity forms, the Environmental Protection Agency (EPA), the Clean Energy Regional Applications Centers, CHP Partnership, developer lists, utility lists, incentive program awardees, industry publications, press releases, and others.

In addition, the Energy Commission provided ICF with CHP sites identified in the *Quarterly Fuels Energy Report* (QFER) that have a capacity of more than 1 MW. The CPUC also provided a list of data on all sizes of CHP systems as reported by the three IOUs in the state of California. Each of the three major utility companies publishes a list of CHP sites that currently have power sales contracts within their QF and Small Generator reports. These lists were all compared to the ICF CHP Installation Database, and during the adjustment process several data corrections have been made and incorporated into the ICF database. These corrections referred to sites listed in other sources as retired being taken out of ICF's list, and sites that are CHP but not listed in ICF being added to the list. Table 8.6 presents the database comparison of the CHP installations as they appear in various sources.

All of the sites in ICF's database that have a capacity greater than 1 MW have been verified as being CHP application through a confirmed source

**TABLE 8.6**
ICF CHP Database Comparison to CEC and CPUC

| Data Source | # Sites | ICF Capacity (MW) | CEC Capacity (MW) | CPUC Capacity (MW) |
|---|---|---|---|---|
| Energy Commission only | 44 | 1545 | 1654 | |
| Energy Commission and CPUC | 131 | 5726 | 5944 | 5694 |
| CPUC only | 164 | 425 | | 431 |
| Utility QF/Small Gen Report | 18 | 2 | | |
| EIA CHP | 18 | 188 | | |
| Unidentified SGIP CHP | 231 | 113 | | |
| Other >1 MW—verified CHP | 72 | 436 | | |
| Other <1 MW—each site not verified | 524 | 82 | | |
| Total | 1202 | 8518 | 7598 | 6125 |

*Source:* Hedman, B., Darrow, K., Wong, E., and Hampson, A. ICF International, 2011. Combined Heat and Power: 2011–2030 Market Assessment. California Energy Commission. CEC-200-2012-002. With permission.

(EIA data, Energy Commission/CPUC lists, SGIP data, utility reports, or various third-party sources). Conversely, the sites under 1 MW were not individually re-verified due to the fact that they do not play an important role. The unidentified SGIP capacity presented in Table 8.6 describes sites that have received SGIP incentives for CHP but are not identified by name in the ICF CHP Installation Database.

### 8.3.1 California Existing CHP Capacity Summary

Approximately 85% of the existing CHP capacity in California resides in large systems with site capacities greater than 20 MW; however these large systems make up only 9% of the number of installations. As shown in Figure 8.1, the largest share of active CHP capacity is present in the industrial sector, while the largest single application is the provision of steam in oil fields for enhanced oil recovery (EOR).

Figure 8.7 shows that the total capacity in the industrial sector is heavily concentrated in six process industries: refining, food processing, pulp and paper, wood products, metals processing, and chemicals. The commercial and institutional sector is represented by a large number of individual market applications, the largest being college/universities, followed by water treatment, healthcare, and government facilities (see Figure 8.8).

The geographic location of the CHP applications in California is spread throughout the utility territories. According to this study, PG&E has the largest share of the CHP capacity in its service area given the concentration

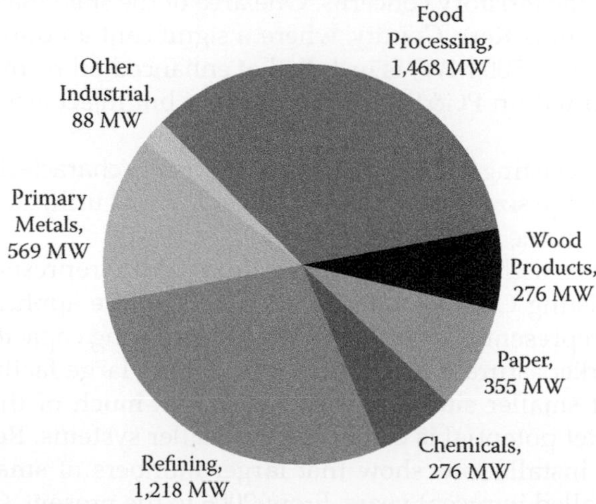


**FIGURE 8.7 (See color insert.)**
Current industrial CHP capacity in California. (From Hedman, B., Darrow, K., Wong, E., and Hampson, A. ICF International, 2011. Combined Heat and Power: 2011–2030 Market Assessment. California Energy Commission. CEC-200-2012-002. With permission.)

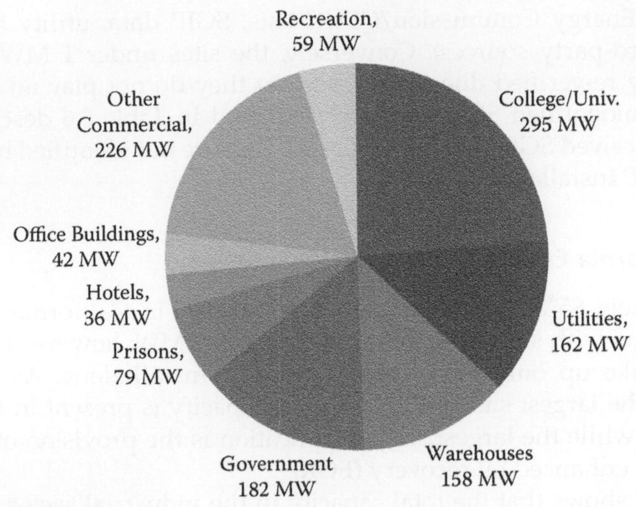


**FIGURE 8.8 (See color insert.)**
Current commercial/institutional CHP capacity in California. (From Hedman, B., Darrow, K., Wong, E., and Hampson, A. ICF International, 2011. Combined Heat and Power: 2011–2030 Market Assessment. California Energy Commission. CEC-200-2012-002. With permission.)

of large oil fields and refineries in its territory. Figure 8.9 presents the distribution of CHP by utility service area. This description depicts the actual physical location of the CHP system and is not important as far as the systems located in one utility territory that sell electricity to other utilities or parties outside the territory concerns. One area of the state that is known to have this problem is Kern County, where a significant amount of the CHP capacity (more than 500 MW) is installed at enhanced oil recovery facilities that are located within PG&E's service territory but inject electricity in the SCE grid.

Generally, the existing CHP installations can also be characterized from the point of view of the size (Figure 8.10), the primary fuel utilized (Figure 8.11), and the type of prime mover (Figure 8.12).

Those systems with a capacity lower than 5 MW represent only 6.2% of the total existing CHP systems in California, while applications larger than 100 MW represent almost 40% of the total existing capacity. As shown below, the market saturation of CHP in the case of large facilities is much higher than at smaller sites. On the other hand, much of the remaining technical market potential is composed of smaller systems. Recent studies carried out in installations show that larger numbers of smaller systems have been installed in recent years. From 2006 to the present, CHP systems with a capacity lower than 5 MW have accounted for 27.7% of the capacity growth.

Figure 8.11 shows that the most important fuel utilized for CHP applications in California is natural gas, which represents 84% of the total installed

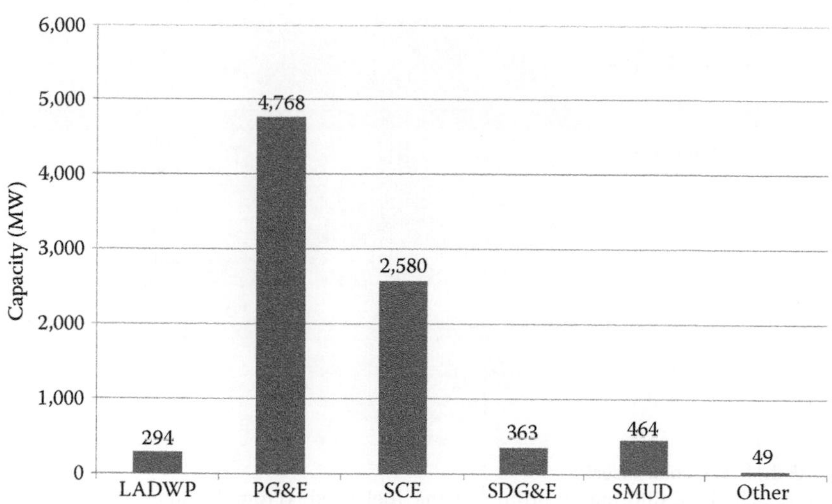


**FIGURE 8.9**
Currently installed CHP in California according to the utility service area. (From Hedman, B., Darrow, K., Wong, E., and Hampson, A. ICF International, 2011. Combined Heat and Power: 2011–2030 Market Assessment. California Energy Commission. CEC-200-2012-002. With permission.)

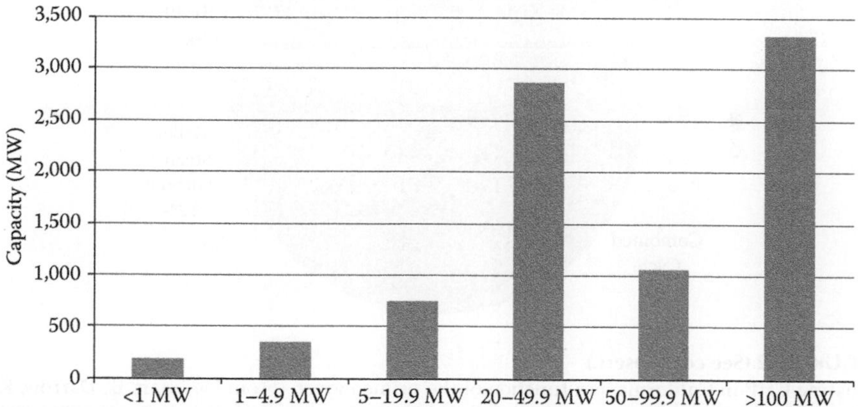


**FIGURE 8.10**
Existing CHP in California according to size. (From Hedman, B., Darrow, K., Wong, E., and Hampson, A. ICF International, 2011. Combined Heat and Power: 2011–2030 Market Assessment. California Energy Commission. CEC-200-2012-002. With permission.)

capacity. In this situation, the coal- and oil-fired systems are becoming rarer, with only eight coal-fired CHP plants, making up 4.5% of capacity, and five oil-fired plants, making up less than one-tenth of 1% of capacity. During the last 5 years, no other new coal- or oil-fired CHP applications have been installed.

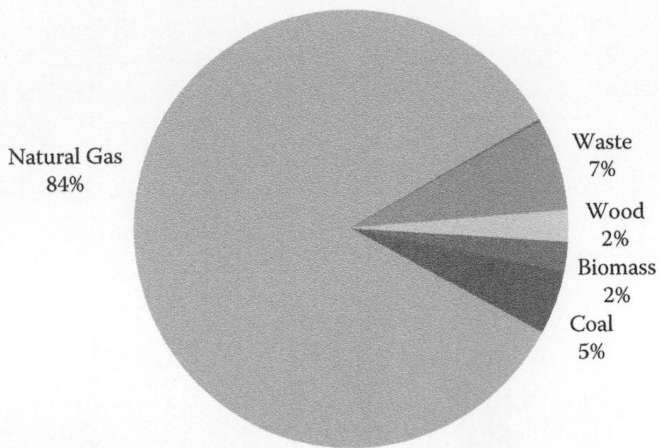


**FIGURE 8.11 (See color insert.)**
Existing CHP in California according to the fuel used. (From Hedman, B., Darrow, K., Wong, E., and Hampson, A. ICF International, 2011. Combined Heat and Power: 2011–2030 Market Assessment. California Energy Commission. CEC-200-2012-002. With permission.)

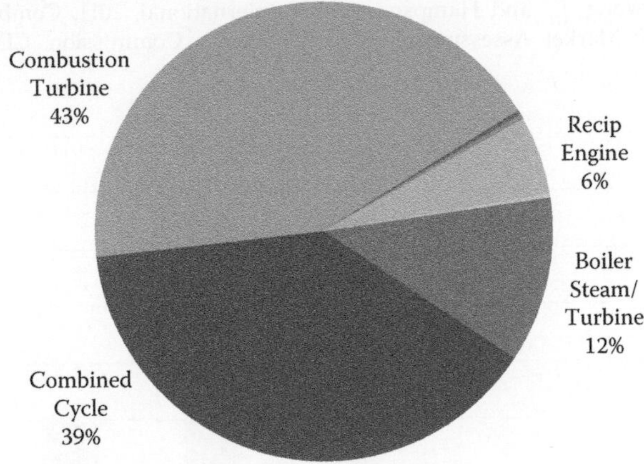


**FIGURE 8.12 (See color insert.)**
Existing CHP in California as a function of the prime mover. (From Hedman, B., Darrow, K., Wong, E., and Hampson, A. ICF International, 2011. Combined Heat and Power: 2011–2030 Market Assessment. California Energy Commission. CEC-200-2012-002. With permission.)

As can be observed in Figure 8.11, the systems based on wood and biomass fuels represent 4.4% of the total capacity, with the bulk of this capacity in the wood products, paper, and food processing industries and in wastewater treatment facilities. The remaining 6.8% is represented by waste fuels primarily from chemical and refining plants.

Given the concentration of large-scale systems in existing CHP systems, the prime movers accounting for the largest capacity are the gas turbines. These usually work in a combined-cycle configuration and have high rated powers. For intermediate sizes, gas turbines on simple-cycle are used. At present, the most common prime mover type in California in terms of number of installations is reciprocating engines; while total capacity is small (5.5%), the reciprocating engine technology represents the greatest number of CHP sites (62%). The emerging technologies, such as MTs and fuel cells, have a small but growing share of the systems. While the amount of capacity provided by MTs and fuel cells remains small (5.6%), in the past 5 years, they are 34% of the number of systems installed (see Figure 8.12).

In addition, California has been negatively influenced by the recent economic crisis. This slowed development of the CHP sector and caused a decrease in CHP capacity due to the fact that industrial or commercial host sites had to shut down. In the last 5 years, there have been 314 MW of CHP applications in California that have ceased to operate because the host facility where they are located has been shut down. It is a good thing that national CHP development trends are starting to change, as the number of CHP systems in the development and construction stage are again increasing.

In order to estimate future CHP development trends, ICF maintains data on CHP systems in the proposed, planning, and construction stages of development. Due to the fact that CHP systems can take many years to install, depending on the system size and host application, tracking systems in development can provide a picture of where the CHP market is heading. The ICF records show that the state of California currently has 11 sites that were installed in 2012, representing 65.1 MW of CHP capacity. This number represents only a fraction of the capacity anticipated to actually enter the market, because in reality, many companies choose not to publicize their CHP development plans.

California is in sixth place in the United States in CHP capacity under development. Other states with large capacities in development are New York, Washington, Michigan, Wisconsin, and Virginia.

## 8.4 CHP Technical Market Potential

In this section, the study provides an estimate of the technical market potential for CHP applications in the commercial/institutional, industrial, and multifamily residential market sectors in California. The technical potential is an estimation of the market size limited only by technology—by the ability of CHP technologies to meet customers' energy needs. The CHP technical potential is calculated considering the CHP electrical capacity that could be installed at existing and new industrial and commercial facilities based on the estimated electric and thermal needs of the site. Technical market potential does not consider evaluation for the economic rate of return, or for other factors such as owner interest in applying CHP, the ability to retrofit, capital availability, variation of energy consumption within customer application/size class, or natural gas availability.

Technical potential plays an important role in understanding the potential size and distribution of the target CHP market in the state. Identifying the technical market potential is the preliminary and probably most important step in assessment of the actual economic market size and ultimate market penetration.

CHP applications are most efficient at sites that have important and simultaneous electric and thermal demands. At industrial facilities, the CHP thermal output has traditionally been in the form of steam used for process heating and for space heating. For commercial and institutional users, thermal output has traditionally been steam or hot water for space heating and potable hot water heating (see also Chapter 7, Case Studies). More recently, the CHP has included the provision of space cooling through the use of absorption chillers, the application being transformed in such a case in CCHP.

Three different types of CHP markets have been included for assessing technical potential:

- Traditional power and heat CHP
- Combined cooling, heating, and power (CCHP)
- Export of power produced by CHP

These first two types of markets were further categorized according to the high load factor and low load factor applications resulting from the analysis of five distinct market segments.

### 8.4.1 Traditional CHP

This market represents CHP applications where the electrical output is used to entirely or partially cover the base load for a facility while thermal energy is used to provide steam or hot water. The most efficient operation for a CHP system is to match the thermal output to base load thermal demand on site. Depending on the type of facility, appropriate sizing can be limited to either electric or thermal. Industrial facilities, on the other hand, often have "excess" thermal load compared to their on-site electric load, which means that the CHP system will generate more power than can be used by the facility.

Commercial facilities instead almost always have excess electric load compared to their thermal load. For a better understanding, two subcategories were considered:

- High load factor systems: The CHP system is operated continuously. This is the case in prisons, colleges, hospitals, etc.
- Low load factor systems: The thermal load is present only 3500 to 5000 hours per year. This is the case in office buildings, health clubs, etc.

### 8.4.2 Combined Cooling Heating and Power (CCHP)

As explained, all or a fraction of the thermal output of a CHP system can be transformed in air conditioning or refrigeration with the addition of an absorption chiller. The advantages are obvious in such a case. A typical CCHP system would supply the facility with the hot water, space heating in the winter months, and finally cooling during the summer months. Two subcategories have been considered:

- High load factor applications: The CCHP system is operated continuously, as in prisons, colleges, hospitals, etc.
- Low load factor applications: Thermal energy is not needed throughout the whole year.

### 8.4.3 CHP Export Market

The previous two categories are based on the limitation that all of the thermal and electric energy must be consumed on site. In the case of many large industrial process facilities, there is often enough steam demand and the thermally dimensioned CHP systems produce more electrical energy than is needed, which could be further injected in the central grid. This "exported" electricity is usually evaluated as a separate market.

### 8.4.4 Technical Potential Methodology

The estimation of the technical market potential consists of the following elements:

- Identifying those applications where the CHP systems provide a reasonable fit to the electric and thermal needs of the consumer
- Quantifying the number and size distribution of the target applications
- Estimating the CHP potential in terms of MW electric capacity
- Subtracting the existing CHP from the identified sites in order to determine the remaining technical potential

### 8.4.5 CHP Target Markets

As a rule, the most efficient and economic CHP operation is attained when: (1) the system is operated at full load most of the time (high load factor application), (2) the thermal output can be 100% consumed on site, and (3) the recovered heat replaces the fuel or electricity purchases.

There are a number of industrial and commercial applications that theoretically have sufficient and simultaneous thermal and electric loads to be covered by CHP systems. Classic examples of this are pulp and paper plants, food processing plants, laundries, health clubs, etc. Most of these commercial and light industrial applications have low base thermal loads when compared to the electric load, but at the same time have high thermal loads in the cooler months for heating. These applications refer to college campuses, hotels, nursing homes, correctional facilities, hospitals, light manufacturing, and so forth.

To identify those applications where the CHP systems provide an efficient covering of electric and thermal consumption, this study analyzed electric and thermal energy (heating and cooling) consumption data from various buildings and industrial sites. The data has also been taken from the Major Industrial Plant Database (MIPD), DOE EIA Commercial Buildings Energy Consumption Survey (CBECS), Commercial Energy Profile Database (CEPD),* the DOE Manufacturing Energy Consumption Survey (MECS), and various market summaries developed by the DOE, the American Gas Association, and the Gas Technology Institute (GTI). Existing CHP installations in the institutional/commercial and industrial sectors were also analyzed to understand the required profile for CHP applications and identify target applications.

National-level data were studied to develop the national average thermal and electric demand profiles according to the application. As stated in the first part of this study, it has been recognized that regional climate and operating factors can influence both thermal and electric load profiles. This is no critical problem for the industrial applications because they tend to be more uniform in their operation nationwide than commercial and institutional facilities. Moreover, commercial facilities use an important share of their purchased energy on heating and cooling, which is further affected by local weather conditions. That is why the sources of electric and thermal load data specific to California have also been reviewed. The CEPD and MIPD facilities in California were studied as well, together with the existing CHP fleet in California. A key role was played by the data source for the commercial sector, the *California Commercial End-Use Survey (CEUS)*. The CEC QFER (Quarterly Fuels Energy Report) data were also used in order to control the amount of energy consumption in the individual applications.

The CHP system design for the three markets previously described is based on the efficient covering of the thermal loads:

- Traditional CHP—Design the CHP system for the base thermal load (domestic hot water, pool heating, etc.), which usually results in a system sized below the base electric load for commercial facilities.
- CCHP—Design the CHP system to include thermally activated cooling to create additional thermal use during the summer months.

* The Commercial Energy Profile Database (CEPD) and Major Industrial Plant Database (MIPD) represent private databases that contain site-specific energy estimates for industrial and commercial facilities. Access in both cases was offered by IHS Inc.

- Export CHP—Design the CHP system to cover the entire thermal load at an industrial site, with excess electricity generation being injected in the main grid. The traditional CHP and the CCHP are based on the assumption that all of the thermal and electric energy is consumed on site.

Tables 8.7 and 8.8 show the CHP market applications classified by these categories as well as their assumed load profiles. It is supposed that those systems having a high load factor operate 7500 hours a year, while the

**TABLE 8.7**

Traditional CHP Target Installations

| NAICS | SIC | Application | Application Type | Load Factor | Export Power Potential |
|---|---|---|---|---|---|
| 311–312 | 20 | Food Processing | Industrial | High | Yes |
| 313 | 22 | Textiles | Industrial | High | Yes |
| 321 | 24 | Lumber and Wood | Industrial | High | Yes |
| 337 | 25 | Furniture | Industrial | High | No |
| 322 | 26 | Paper | Industrial | High | Yes |
| 325 | 28 | Chemicals | Industrial | High | Yes |
| 324 | 29 | Petroleum Refining | Industrial | High | Yes |
| 326 | 30 | Rubber/Miscellaneous Plastics | Industrial | High | No |
| 331 | 33 | Primary Metals | Industrial | High | No |
| 332 | 34 | Fabricated Metals | Industrial | High | No |
| 333 | 35 | Machinery/Computer Equipment | Industrial | High | No |
| 336 | 37 | Transportation Equipment | Industrial | High | No |
| 335 | 38 | Instruments | Industrial | High | No |
| 339 | 39 | Miscellaneous Manufacturing | Industrial | High | Yes |
| 2213 | 4941 | Water Treatment/Sanitary | Commercial/ Institutional | High | No |
| 92214 | 9223 | Prisons | Commercial/ Institutional | High | No |
| 8123 | 7211 | Laundries | Commercial/ Institutional | Low | No |
| 71394 | 7991 | Health Clubs | Commercial/ Institutional | Low | No |
| 71391 | 7992 | Golf/Country Clubs | Commercial/ Institutional | Low | No |
| 8111 | 7542 | Car Washes | Commercial/ Institutional | Low | No |

*Source:* Hedman, B., Darrow, K., Wong, E., and Hampson, A. ICF International, 2011. Combined Heat and Power: 2011–2030 Market Assessment. California Energy Commission. CEC-200-2012-002. With permission.

**TABLE 8.8**

CCHP and the Power Target Installations

| NAICS | SIC | Application | Application Type | Load Factor |
|---|---|---|---|---|
| 531 | 6513 | Apartments | Commercial/Institutional | High |
| 721 | 7011 | Hotels | Commercial/Institutional | High |
| 623 | 8051 | Nursing Homes | Commercial/Institutional | High |
| 622 | 8062 | Hospitals | Commercial/Institutional | High |
| 6113 | 8221 | Colleges/Universities | Commercial/Institutional | High |
| 518 | 7374 | Data Centers | Commercial/Institutional | High |
| 531 | 6512 | Community Office Buildings | Commercial/Institutional | Low |
| 6111 | 8211 | Schools | Commercial/Institutional | Low |
| 612 | 8412 | Museums | Commercial/Institutional | Low |
| 491 | 43 | Post Offices | Commercial/Institutional | Low |
| 452 | 50 | Big Box Retail | Commercial/Institutional | Low |
| 48811 | 4581 | Airport Facilities | Commercial/Institutional | Low |
| 445 | 5411 | Food Sales | Commercial/Institutional | Low |
| 722 | 5812 | Restaurants | Commercial/Institutional | Low |
| 512131 | 7832 | Movie Theaters | Commercial/Institutional | Low |
| 92 | 9100 | Government Buildings | Commercial/Institutional | Low |

*Source:* Hedman, B., Darrow, K., Wong, E., and Hampson, A. ICF International, 2011. Combined Heat and Power: 2011–2030 Market Assessment. California Energy Commission. CEC-200-2012-002. With permission.

applications with a low load factor operate 5000 hours a year. The load profile together with the category determine the four markets defined above. Each of these applications is presented with both the corresponding North American Industry Classification System (NAICS) code and Standard Industrial Classification (SIC) code.

## 8.5 California Target CHP Facilities

Various industrial and commercial facility databases have been employed to identify the number of target application facilities in California by sector and by size (electric demand) which comply with the thermal and electric load requirements for CHP systems. The most important data source to identify potential targets for CHP installations in California was the Dun & Bradstreet (D&B) *Hoovers* Database. This database was acquired in October 2011 and contains information about the majority of businesses throughout the country. It can be sorted to provide a listing of industrial and commercial facilities in a specific region. This study used a set of data consisting of facilities in California that have more than five employees and are in the target applications specified above. The site data includes the following information:

- Name of the company
- Facility location
- Type of business (primary SIC code and primary NAICS code)
- Number of employees (at total company and at individual site)
- Annual sales
- Facility size (in square-feet)

Almost 50,000 facilities from the D&B Hoovers database, including 14,630 industrial* sites and 35,310 commercial sites, were analyzed for CHP potential in this study. Other industrial facilities from other sources were employed to complete the D&B *Hoovers* as far as the large industrial market segment is concerned. The large refineries have been carefully reviewed to make sure that the estimates for additional CHP potential were consistent with current refining industry assumptions. In a similar study entitled "ICF 2009 CHP Market Assessment for California,"† a list of the major refineries in California was compiled, containing detailed information on their electric demand and process steam flows. This information was used further to independently calculate the remaining potential for CHP in the refining sector.

The large industrial plants in the combined list were also independently checked to compare the electric and boiler fuel data, and the estimated values were calculated through the methodology detailed below.

### 8.5.1 Quantify Electric and Thermal Loads for CHP Target Applications

To obtain an estimate for the total technical CHP potential in California, a hypothetical CHP has been attributed to each of the target facilities system which is designed in accordance with its thermal and electrical loads. In this case, the sum of all the individual CHP system capacities would then result in the overall total CHP potential for the state.

### 8.5.2 Electric Load Estimation

It has been assumed that the CHP systems would be sized to comply with the base thermal loads (cooling and heating) of a site unless the CHP system design exceeded the average facility electric demand. In this situation,

---

* All the sites listed in the D&B *Hoovers* database are categorized according to their respective market applications based on the primary NAICS code listed in the database.

† For more information, please refer to California Energy Commission, Public Interest Energy Research Program. Combined Heat and Power Market Assessment. Prepared by ICF International, Inc. February 2010.

it is supposed that the industrial sites will export the excess electricity to the central grid, while the commercial sites would limit the system size to the average electric demand of the facility. The total electrical load measured in kWh is estimated for each facility through algorithms from the CHP Market Model based on characteristics such as annual sales, surface, or number of employees. The average electric consumption of each facility was estimated by dividing the total kWh electricity load to the typical operating hours corresponding to the load factor of the application (5000 hours a year for low load factor and 7000 hours a year for a high load factor).

From the 50,000 sites in California analyzed for CHP potential, almost half were removed from the analysis due to the fact that the estimated electric demand was missing. This estimation required a minimum electric demand of 50 kW for a facility that had to be included in the technical potential. After analyzing this minimum electric demand, approximately 25,000 sites remained as potential CHP candidates.

### 8.5.3 Thermal Load Estimation

As described, this study supposes that the CHP applications are designed to comply with the base thermal loads (cooling and heating) of each site. The estimation of the thermal load is of ground zero importance to properly project the CHP system for high thermal utilization and to determine whether the thermal load would limit the CHP system size.

To optimize the thermal demand estimates for the commercial sector, the CEUS was used to corroborate the thermal demand estimates with the Californian climate. This data source supplied enough information about the end-use energy consumption in commercial and industrial facilities so that the average power-to-heat ratio factors for each target site could be developed.

In the present study the estimation methodology has changed compared to the ICF's 2009 assessment of CHP potential in California.* The detailed electric and thermal data available now were used to develop size-specific thermal factors for each CHP target application that are used to estimate the CHP system size as a function of average electric consumption. The thermal factor is based on both the power-to-heat (P/H) ratio of a typical CHP system for that application and the (P/H) ratio of the application.

### 8.5.4 CHP System Sizing

The thermal and electrical data described above were used to develop thermal factors for each application, which are used to estimate the CHP system

* For more information, please refer to California Energy Commission, Public Interest Energy Research Program. Combined Heat and Power Market Assessment. Prepared by ICF International, Inc. February 2010.

size for each potential site as a function of average electric consumption. The thermal factor is based on both the P/H ratio of the application as well as the P/H ratio of a typical CHP system for that application. This thermal factor is then multiplied by the average electric demand to determine the estimated CHP system size for each site. When the thermal factor equals one, this would mean that the CHP system capacity is equal to the average electric consumption of the facility. When the value is less than one, this would indicate that the application is thermally limited, and the resulting CHP system size would be below the average electric consumption of the site. When the thermal factor is greater than one, this indicates that the CHP system is designed to cover the thermal load and would produce more electricity than can be consumed on site, resulting in excess power that could be injected in the central network. A certain number of industrial applications in California have thermal factors greater than one, indicating the capacity to export power to the grid for CHP systems is sized to meet thermal loads.

After the potential CHP capacity has been determined for each of the potential facilities, the existing CHP installations in California were corroborated to the list and subtracted from the CHP technical potential. If a site with an existing CHP application had a higher amount of technical potential than is currently installed, the difference was considered to represent remaining potential at the site.

### 8.5.5 Technical Potential Results

The same methodology has been employed to assess the estimates for CHP technical market potential for both existing facilities in 2011 and new facility growth between 2011 and 2030.

The estimates are categorized into the CHP technical potential which supplies the on-site electric demands at target facilities and additional CHP technical potential which is available if the facilities are allowed to sell electricity to the grid (export capacity). This excess CHP capacity is presented in the export tables.

The total technical market potential (export and on-site) for CHP equals 14,293 MW in 2011 in the case of the potential at existing commercial and industrial facilities, while another 1671 MW are expected from new or improved commercial and industrial facilities during the period 2011–2030. The forecast for 2030 is almost 16,000 MW.

#### *8.5.5.1 Technical Potential—2011*

Table 8.9 presents the number of potential facilities supplied by the utility companies. As can be seen, the two regions with the largest amount of technical potential are SCE and PG&E. These two utilities cover large geographic areas. Since PG&E also has the largest number of existing CHP installations, the remaining CHP potential indicates that SCE has more room for growth

**TABLE 8.9**

CHP Technical Potential (MW) as a Function of the Utility Region in 2011

| Utility Region | 50–500 kW | 500–1000 kW | 1–5 MW | 5–20 MW | >20 MW | Total |
|---|---|---|---|---|---|---|
| LADWP | 229 | 189 | 299 | 197 | 179 | 1093 |
| PG&E | 1033 | 435 | 998 | 591 | 297 | 3354 |
| SCE | 1040 | 385 | 942 | 604 | 289 | 3259 |
| SDG&E | 220 | 105 | 212 | 109 | 46 | 692 |
| SMUD | 81 | 43 | 98 | 84 | 21 | 328 |
| Other North | 57 | 23 | 45 | 72 | 0 | 196 |
| Other South | 106 | 41 | 99 | 90 | 0 | 336 |
| Total (MW) | 2765 | 1221 | 2693 | 1747 | 833 | 9259 |

*Source:* Hedman, B., Darrow, K., Wong, E., and Hampson, A. ICF International, 2011. Combined Heat and Power: 2011–2030 Market Assessment. California Energy Commission. CEC-200-2012-002. With permission.

in CHP capacity as a percentage of current CHP installations. Also, the Los Angeles Department of Water and Power (LADWP) has significant potential, given the small size of its service area.

Tables 8.10 through 8.14 summarize the present technical potential estimates categorized according to application, size, and utility territory. The technical potential for CHP applications is higher in industrial sectors, which now have an important number of existing CHP installations such as food processing, chemicals, and paper production. Because many of the large industrial facilities in California already have CHP systems, the majority of the potential now is represented by systems with a capacity in the range between 1 MW and 20 MW.

The CHP potential for commercial facilities is heavily concentrated in size ranges below 5 MW, where approximately 75% of the technical potential lies. This potential is stimulated by several large applications that incorporate cooling into the CHP system design, including government buildings, colleges/universities, commercial buildings, and so forth. The estimation of the CHP export market is based above all on the excess power capacity at the largest 100 industrial facilities in the state of California, characterized from the point of view of the steam demand. An important percentage of this potential comes from a few large chemical plants, refineries, or food processing facilities. The forecast of the technical potential for additional export CHP capacity in enhanced oil recovery applications is based on a 1999 EPRI study analyzing the potential at 10 existing oil fields and the degree of market saturation that already exists for CHP.* These forecasts were increased by 26% to reflect the improved levels of Enhanced Oil Recovery (EOR) steam injection as described in the 2000 through 2010

* For more information, please refer to Enhanced Oil Recovery Scoping Study, EPRI, Palo Alto, CA: 1999. TR-113836, http://www.energy.ca.gov/process/pubs/electrotech_opps_tr113836.pdf.

**TABLE 8.10**

CHP Technical Potential at Existing Industrial Facilities in 2011

| NAICS | Application | 50–500 kW (MW) | 500–1000 MW (MW) | 1–5 MW (MW) | 5–20 MW (MW) | >20 MW (MW) | Total (MW) |
|---|---|---|---|---|---|---|---|
| 311 | Food | 226 | 109 | 258 | 196 | 56 | 845 |
| 313 | Textiles | 45 | 10 | 30 | 8 | 26 | 119 |
| 321 | Lumber and Wood | 56 | 17 | 45 | 23 | 25 | 165 |
| 337 | Furniture | 0 | 0 | 0 | 0 | 0 | 0 |
| 322 | Paper | 61 | 54 | 168 | 132 | 20 | 434 |
| 323 | Printing | 0 | 0 | 3 | 0 | 0 | 3 |
| 325 | Chemicals | 149 | 99 | 396 | 360 | 97 | 1100 |
| 324 | Petroleum Refining | 11 | 30 | 62 | 58 | 125 | 285 |
| 326 | Rubber/Miscellaneous Plastics | 44 | 18 | 17 | 6 | 0 | 86 |
| 327 | Stone/Clay/Glass | 12 | 12 | 23 | 0 | 0 | 47 |
| 331 | Primary Metals | 28 | 5 | 13 | 9 | 0 | 55 |
| 332 | Fabricated Metals | 14 | 3 | 1 | 0 | 0 | 18 |
| 333 | Machinery/Computer Equipment | 10 | 5 | 10 | 0 | 0 | 25 |
| 336 | Transportation Equipment | 18 | 13 | 15 | 26 | 0 | 73 |
| 335 | Instruments | 13 | 1 | 3 | 0 | 37 | 53 |
| 339 | Miscellaneous Manufacturing | 0 | 0 | 0 | 0 | 0 | 0 |
| | Total (MW) | 688 | 375 | 1042 | 818 | 385 | 3309 |

*Source:* Hedman, B., Darrow, K., Wong, E., and Hampson, A. ICF International, 2011. Combined Heat and Power: 2011–2030 Market Assessment. California Energy Commission. CEC-200-2012-002. With permission.

annual reports from the Division of Oil, Gas, and Geothermal Resources (Department of Conservation).

The total technical CHP export potential is valued at 5034 MW. This export potential is geographically located in this study for placement in utility service territories. Facilities capable of exporting power may choose to sell their electricity to any entity they wish, including those located outside their geographic area.

Table 8.13 presents the export CHP technical potential according to utility area. As demonstrated in this table, the utility with the largest amount of CHP export technical potential is PG&E due to the important presence of EOR opportunities in the PG&E service territory. In Table 8.14, the total technical potential for CHP in California in 2011 is presented according to the CHP market sector. Table 8.14 indicates that there is more remaining potential in commercial facilities than in the industrial facilities, which is an important difference compared to the traditional characterization

**TABLE 8.11**

CHP Technical Potential at Existing Commercial Facilities in 2011

| NAICS | Application | 50–500 kW (MW) | 500–1000 MW (MW) | 1–5 MW (MW) | 5–20 MW (MW) | >20 MW (MW) | Total (MW) |
|---|---|---|---|---|---|---|---|
| 491 | Post Offices | 7 | 2 | 0 | 0 | 0 | 9 |
| 452 | Retail | 245 | 36 | 15 | 0 | 0 | 296 |
| 493 | Refrigerated Warehouses | 16 | 6 | 4 | 5 | 0 | 31 |
| 48811 | Airports | 1 | 2 | 8 | 29 | 27 | 67 |
| 2213 | Water Treatment | 28 | 7 | 7 | 0 | 0 | 41 |
| 445 | Food Stores | 220 | 8 | 8 | 0 | 0 | 235 |
| 722 | Restaurants | 163 | 9 | 7 | 9 | 0 | 187 |
| 531 | Commercial Buildings | 294 | 368 | 511 | 0 | 0 | 1172 |
| 531 | Multi-family Buildings | 105 | 111 | 72 | 0 | 0 | 288 |
| 721 | Hotels | 166 | 76 | 158 | 38 | 0 | 439 |
| 8123 | Laundries | 25 | 4 | 2 | 0 | 0 | 31 |
| 518 | Data Centers | 19 | 6 | 7 | 0 | 0 | 32 |
| 8111 | Car Washes | 18 | 1 | 0 | 0 | 0 | 18 |
| 512131 | Movie Theaters | 1 | 0 | 1 | 0 | 0 | 2 |
| 71394 | Health Clubs | 55 | 6 | 3 | 0 | 0 | 63 |
| 71391 | Golf/Country Clubs | 63 | 1 | 2 | 0 | 0 | 66 |
| 623 | Nursing Homes | 128 | 4 | 14 | 0 | 0 | 146 |
| 622 | Hospitals | 54 | 56 | 267 | 58 | 0 | 435 |
| 6111 | Schools | 216 | 23 | 32 | 9 | 0 | 280 |
| 6113 | College/University | 50 | 24 | 229 | 649 | 396 | 1348 |
| 612 | Museums | 9 | 1 | 0 | 0 | 0 | 11 |
| 91 | Government Buildings | 182 | 92 | 268 | 131 | 25 | 698 |
| 92214 | Prisons | 12 | 5 | 35 | 0 | 0 | 52 |
| | Total (MW) | 2077 | 846 | 1650 | 929 | 447 | 5950 |

*Source:* Hedman, B., Darrow, K., Wong, E., and Hampson, A. ICF International, 2011. Combined Heat and Power: 2011–2030 Market Assessment. California Energy Commission. CEC-200-2012-002. With permission.

of CHP target markets. Another conclusion presented in this table is that there is a concentration of potential at the low capacity sites, meaning that many large facilities already have CHP systems for their on-site needs.

In addition to the technical potential figures estimated using ICF's standard methodology, a high electric focus case was calculated to measure the increase in potential that could be achieved if electric utilities would owe large CHP systems and would design these to maximize power production. In the standard methodology, the large industrial sites with high thermal and electric loads have their CHP technical potential estimated assuming

**TABLE 8.12**

CHP Export Potential at Existing Industrial Facilities in 2011

| NAICS | Application | 50–500 kW (MW) | 500–1000 MW (MW) | 1–5 MW (MW) | 5–20 MW (MW) | >20 MW (MW) | Total (MW) |
|---|---|---|---|---|---|---|---|
| 211 | Enhanced Oil Recovery | 0 | 0 | 0 | 0 | 1350 | 1350 |
| 311 | Food | 0 | 0 | 91 | 97 | 297 | 486 |
| 313 | Textiles | 0 | 0 | 0 | 9 | 4 | 12 |
| 321 | Lumber and Wood | 0 | 0 | 38 | 31 | 106 | 175 |
| 337 | Furniture | 0 | 0 | 0 | 0 | 0 | 0 |
| 322 | Paper | 0 | 0 | 24 | 329 | 601 | 955 |
| 323 | Printing | 0 | 0 | 0 | 10 | 0 | 10 |
| 325 | Chemicals | 0 | 0 | 89 | 267 | 543 | 899 |
| 324 | Petroleum Refining | 0 | 0 | 43 | 95 | 946 | 1084 |
| 326 | Rubber/Miscellaneous Plastics | 0 | 0 | 0 | 12 | 0 | 12 |
| 327 | Stone/Clay/Glass | 0 | 0 | 0 | 0 | 0 | 0 |
| 331 | Primary Metals | 0 | 0 | 0 | 8 | 0 | 8 |
| 332 | Fabricated Metals | 0 | 0 | 0 | 10 | 0 | 10 |
| 333 | Machinery/Computer Equipment | 0 | 0 | 0 | 0 | 0 | 0 |
| 336 | Transportation Equipment | 0 | 0 | 0 | 27 | 0 | 27 |
| 335 | Instruments | 0 | 0 | 0 | 5 | 0 | 5 |
| 339 | Miscellaneous Manufacturing | 0 | 0 | 0 | 0 | 0 | 0 |
| | Total (MW) | 0 | 0 | 286 | 901 | 3847 | 5034 |

*Source:* Hedman, B., Darrow, K., Wong, E., and Hampson, A. ICF International, 2011. Combined Heat and Power: 2011–2030 Market Assessment. California Energy Commission. CEC-200-2012-002. With permission.

**TABLE 8.13**

CHP Export Potential in 2011 as a Function of the Utility Territory

| Utility Region | 50–500 kW (MW) | 500–1000 MW (MW) | 1–5 MW (MW) | 5–20 MW (MW) | >20 MW (MW) | Total (MW) |
|---|---|---|---|---|---|---|
| LADWP | 0 | 0 | 5 | 34 | 240 | 279 |
| PG&E | 0 | 0 | 126 | 322 | 2640 | 3088 |
| SCE | 0 | 0 | 105 | 433 | 691 | 1229 |
| SDG&E | 0 | 0 | 10 | 25 | 171 | 206 |
| SMUD | 0 | 0 | 5 | 32 | 0 | 37 |
| Other North | 0 | 0 | 19 | 13 | 106 | 138 |
| Other South | 0 | 0 | 16 | 42 | 0 | 58 |
| Total (MW) | 0 | 0 | 286 | 901 | 3847 | 5034 |

*Source:* Hedman, B., Darrow, K., Wong, E., and Hampson, A. ICF International, 2011. Combined Heat and Power: 2011–2030 Market Assessment. California Energy Commission. CEC-200-2012-002. With permission.

**TABLE 8.14**

Total CHP Technical Potential at Existing Facilities—Industrial and Commercial—In 2011 by CHP Market Share

| Market Type | 50–500 kW (MW) | 500–1000 MW (MW) | 1–5 MW (MW) | 5–20 MW (MW) | >20 MW (MW) | Total (MW) |
|---|---|---|---|---|---|---|
| Industrial On-site | 688 | 375 | 1,042 | 818 | 385 | 3,309 |
| Commercial—Traditional | 200 | 23 | 49 | 0 | 0 | 272 |
| Commercial—Heating & Cooling | 1,773 | 712 | 1,529 | 929 | 447 | 5,390 |
| Residential—Heating & Cooling | 105 | 111 | 72 | 0 | 0 | 288 |
| Export Existing | 0 | 0 | 286 | 901 | 3,847 | 5,034 |
| Total (MW) | 2,765 | 1,221 | 2,978 | 2,648 | 4,679 | 14,293 |

*Source:* Hedman, B., Darrow, K., Wong, E., and Hampson, A. ICF International, 2011. Combined Heat and Power: 2011–2030 Market Assessment. California Energy Commission. CEC-200-2012-002. With permission.

**TABLE 8.15**

CHP Export Potential According to the High Electric Focus Case

| Utility Region | 50–500 kW (MW) | 500–1000 MW (MW) | 1–5 MW (MW) | 5–20 MW (MW) | >20 MW (MW) | Total (MW) |
|---|---|---|---|---|---|---|
| LADWP | 0 | 0 | 5 | 34 | 592 | 631 |
| PG&E | 0 | 0 | 126 | 322 | 2876 | 3323 |
| SCE | 0 | 0 | 105 | 433 | 1425 | 1963 |
| SDG&E | 0 | 0 | 10 | 25 | 330 | 365 |
| SMUD | 0 | 0 | 5 | 32 | 0 | 37 |
| Other North | 0 | 0 | 19 | 13 | 195 | 228 |
| Other South | 0 | 0 | 16 | 42 | 0 | 58 |
| Total (MW) | 0 | 0 | 286 | 901 | 5419 | 6606 |

*Source:* Hedman, B., Darrow, K., Wong, E., and Hampson, A. ICF International, 2011. Combined Heat and Power: 2011–2030 Market Assessment. California Energy Commission. CEC-200-2012-002. With permission.

they would install a simple cycle gas turbine. For High Electric Focus cases, it is supposed that these large industrial sites with technical potential over 50 MW would alternatively install combined cycle systems, which have higher P/H ratios, and would thus generate greater amounts of electricity output.

Table 8.15 presents an increased export capacity available supposing that combined cycle systems would be installed at sites with important technical potential.

### 8.5.6 Technical Potential Growth between 2011 and 2030

The 2011 technical potential forecast is based on the facility data in the potential CHP site list, while the 2030 estimate includes economic growth predictions for target applications between 2011 and 2030. In order to obtain an accurate estimate for the development of new industrial and commercial facilities and for the improvement in existing facilities between 2011 and 2030, economic forecasts regarding growth by target market applications in California were analyzed. Tables 8.16 and 8.17 show the growth factors used in the analysis for the interval 2011–2030 according to each individual sector.

These growth predictions are based on the EIA's (Energy Information Administration's) *Annual Energy Outlook (AEO) 2011* Reference Case, which contains the expected growth rates by industry application up to 2030.

Tables 8.18 and 8.19 describe additional CHP technical market potential given the forecasted economic growth in California over the time period

**TABLE 8.16**

Current Industrial Application Growth Projections

| Application | 2011–2030 Growth Rate,% |
|---|---|
| Food | 18.98 |
| Textiles | 0.00 |
| Lumber and Wood | 11.10 |
| Furniture | 11.10 |
| Paper | 6.07 |
| Publishing | 0.00 |
| Chemicals | 0.00 |
| Petroleum Refining | 0.00 |
| Rubber/Miscellaneous Plastics | 0.00 |
| Stone/Clay/Glass | 0.00 |
| Primary Metals | 0.00 |
| Fabricated Metals | 13.48 |
| Machinery/Computer Equipment | 13.48 |
| Transportation Equipment | 13.48 |
| Instruments | 13.48 |
| Miscellaneous Manufacturing | 10.09 |

*Source:* U.S. Energy Information Administration (EIA), *2011 Annual Energy Outlook*, Reference Case, Washington, DC.

**TABLE 8.17**

Commercial Application Growth Projections

| Application | 2011–2030 Growth Rate,% |
|---|---|
| Post Offices | 12.11 |
| Big Box Retail | 28.10 |
| Warehouses | 15.91 |
| Airport Facilities | 26.79 |
| Wastewater Treatment/Sanitary | 24.23 |
| Food Stores | 21.43 |
| Restaurants | 20.00 |
| Commercial Office Buildings | 24.23 |
| Apartments | 11.10 |
| Hotels | 26.79 |
| Laundries | 26.79 |
| Data Centers | 24.23 |
| Car Washes | 24.23 |
| Movie Theaters | 28.10 |
| Health Clubs | 24.23 |
| Golf/Country Clubs | 26.79 |
| Nursing Homes | 30.61 |
| Hospitals | 30.61 |
| Schools | 12.77 |
| Colleges/Universities | 12.77 |
| Museums | 14.81 |
| Government Buildings | 24.23 |
| Prisons | 26.79 |

*Source:* U.S. Energy Information Administration (EIA), *2011 Annual Energy Outlook*, Reference Case, Washington, DC.

**TABLE 8.18**

Total CHP Technical Potential Growth between 2011 and 2030 According to CHP Market Share

| Market Type | 50–500 kW (MW) | 500–1 MW (MW) | 1–5 MW (MW) | 5–20 MW (MW) | >20 MW (MW) | Total (MW) |
|---|---|---|---|---|---|---|
| Industrial On-site | 60 | 29 | 68 | 51 | 20 | 228 |
| Commercial—Traditional | 51 | 6 | 13 | 0 | 0 | 70 |
| Commercial—Heating & Cooling | 408 | 173 | 363 | 154 | 64 | 1162 |
| Residential—Heating & Cooling | 12 | 12 | 8 | 0 | 0 | 32 |
| Export Existing | 0 | 0 | 9 | 40 | 131 | 180 |
| Total (MW) | 531 | 220 | 461 | 245 | 214 | 1671 |

*Source:* Hedman, B., Darrow, K., Wong, E., and Hampson, A. ICF International, 2011. Combined Heat and Power: 2011–2030 Market Assessment. California Energy Commission. CEC-200-2012-002. With permission.

**TABLE 8.19**

Total Industrial CHP Potential in 2030

| NAICS | Application | 50–500 kW (MW) | 500–1 MW (MW) | 1–5 MW (MW) | 5–20 MW (MW) | >20 MW (MW) | Total (MW) |
|---|---|---|---|---|---|---|---|
| 311 | Food | 269 | 129 | 307 | 233 | 67 | 1005 |
| 313 | Textiles | 45 | 10 | 30 | 8 | 26 | 119 |
| 321 | Lumber and Wood | 62 | 19 | 50 | 25 | 28 | 184 |
| 337 | Furniture | 0 | 0 | 0 | 0 | 0 | 0 |
| 322 | Paper | 65 | 57 | 178 | 140 | 21 | 461 |
| 323 | Printing | 0 | 0 | 3 | 0 | 0 | 3 |
| 325 | Chemicals | 149 | 99 | 396 | 360 | 97 | 1100 |
| 324 | Petroleum Refining | 11 | 30 | 62 | 58 | 125 | 285 |
| 326 | Rubber/Miscellaneous Plastics | 44 | 18 | 17 | 6 | 0 | 86 |
| 327 | Stone/Clay/Glass | 12 | 12 | 23 | 0 | 0 | 47 |
| 331 | Primary Metals | 28 | 5 | 13 | 9 | 0 | 55 |
| 332 | Fabricated Metals | 16 | 3 | 1 | 0 | 0 | 20 |
| 333 | Machinery/Computer Equipment | 12 | 6 | 11 | 0 | 0 | 29 |
| 336 | Transportation Equipment | 21 | 15 | 18 | 30 | 0 | 83 |
| 335 | Instruments | 14 | 1 | 3 | 0 | 41 | 60 |
| 339 | Miscellaneous Manufacturing | 0 | 0 | 0 | 0 | 0 | 0 |
| | Total (MW) | 748 | 404 | 1110 | 869 | 405 | 3537 |

*Source:* Hedman, B., Darrow, K., Wong, E., and Hampson, A. ICF International, 2011. Combined Heat and Power: 2011–2030 Market Assessment. California Energy Commission. CEC-200-2012-002. With permission.

of the study. Tables 8.20 and 8.21 present the predicted total commercial CHP potential in 2030 and the total predicted export potential in 2030, respectively. The total technical potential for CHP in 2030 represents the summation between the 2011 technical potential and the projected growth in CHP potential in the interval 2011–2030. Table 8.19 summarizes the total technical potential for CHP in 2030. Table 8.22 depicts the total technical potential for CHP in 2030 according to the utility territory. Figure 8.13 presents the existing CHP capacity and the remaining CHP potential (up to 2030) according to the utility service area. The most important regions as far as growth is concerned are in the SCE and PG&E service territories. However, both SDG&E® and LADWP® show significant room for growth in CHP capacity.

**TABLE 8.20**

Total Commercial CHP Potential in 2030

| NAICS | Application | 50–500 kW (MW) | 500–1 MW (MW) | 1–5 MW (MW) | 5–20 MW (MW) | >20 MW (MW) | Total (MW) |
|---|---|---|---|---|---|---|---|
| 491 | Post Offices | 8 | 2 | 0 | 0 | 0 | 10 |
| 452 | Retail | 314 | 46 | 19 | 0 | 0 | 379 |
| 493 | Refrigerated Warehouses | 19 | 7 | 5 | 6 | 0 | 36 |
| 48811 | Airports | 1 | 2 | 10 | 37 | 34 | 85 |
| 2213 | Water Treatment | 35 | 9 | 9 | 0 | 0 | 52 |
| 445 | Food Stores | 267 | 10 | 10 | 0 | 0 | 286 |
| 722 | Restaurants | 196 | 11 | 8 | 11 | 0 | 225 |
| 531 | Commercial Buildings | 365 | 457 | 635 | 0 | 0 | 1,457 |
| 531 | Multifamily Buildings | 117 | 123 | 80 | 0 | 0 | 320 |
| 721 | Hotels | 210 | 96 | 200 | 48 | 0 | 556 |
| 8123 | Laundries | 32 | 5 | 3 | 0 | 0 | 39 |
| 518 | Data Centers | 24 | 7 | 9 | 0 | 0 | 40 |
| 8111 | Car Washes | 22 | 1 | 0 | 0 | 0 | 23 |
| 512131 | Movie Theaters | 1 | 0 | 1 | 0 | 0 | 3 |
| 71394 | Health Clubs | 68 | 7 | 4 | 0 | 0 | 79 |
| 71391 | Golf/Country Clubs | 80 | 1 | 3 | 0 | 0 | 84 |
| 623 | Nursing Homes | 167 | 5 | 18 | 0 | 0 | 191 |
| 622 | Hospitals | 70 | 73 | 349 | 76 | 0 | 568 |
| 6111 | Schools | 244 | 26 | 36 | 10 | 0 | 316 |
| 6113 | College/University | 56 | 27 | 258 | 732 | 447 | 1,520 |
| 612 | Museums | 10 | 1 | 0 | 0 | 0 | 12 |
| 91 | Government Buildings | 226 | 114 | 333 | 163 | 31 | 867 |
| 92214 | Prisons | 15 | 6 | 44 | 0 | 0 | 66 |
| | Total (MW) | 2,548 | 1,039 | 2,034 | 1,082 | 512 | 7,214 |

*Source:* Hedman, B., Darrow, K., Wong, E., and Hampson, A. ICF International, 2011. Combined Heat and Power: 2011–2030 Market Assessment. California Energy Commission. CEC-200-2012-002. With permission.

**TABLE 8.21**
Total Export CHP Potential in 2030

| NAICS | Application | 50–500 kW (MW) | 500–1 MW (MW) | 1–5 MW (MW) | 5–20 MW (MW) | >20 MW (MW) | Total (MW) |
|---|---|---|---|---|---|---|---|
| 211 | Enhanced Oil Recovery | 0 | 0 | 0 | 0 | 1350 | 1350 |
| 311 | Food | 0 | 0 | 106 | 103 | 370 | 579 |
| 313 | Textiles | 0 | 0 | 0 | 9 | 4 | 12 |
| 321 | Lumber and Wood | 0 | 0 | 39 | 35 | 120 | 195 |
| 337 | Furniture | 0 | 0 | 0 | 0 | 0 | 0 |
| 322 | Paper | 0 | 0 | 24 | 351 | 645 | 1020 |
| 323 | Printing | 0 | 0 | 0 | 10 | 0 | 10 |
| 325 | Chemicals | 0 | 0 | 89 | 267 | 543 | 899 |
| 324 | Petroleum Refining | 0 | 0 | 43 | 95 | 946 | 1084 |
| 326 | Rubber/Miscellaneous Plastics | 0 | 0 | 0 | 12 | 0 | 12 |
| 327 | Stone/Clay/Glass | 0 | 0 | 0 | 0 | 0 | 0 |
| 331 | Primary Metals | 0 | 0 | 0 | 8 | 0 | 8 |
| 332 | Fabricated Metals | 0 | 0 | 0 | 12 | 0 | 12 |
| 333 | Machinery/Computer Equipment | 0 | 0 | 0 | 0 | 0 | 0 |
| 336 | Transportation Equipment | 0 | 0 | 0 | 32 | 0 | 32 |
| 335 | Instruments | 0 | 0 | 0 | 6 | 0 | 6 |
| 339 | Miscellaneous Manufacturing | 0 | 0 | 0 | 0 | 0 | 0 |
| | Total (MW) | 0 | 0 | 302 | 939 | 3978 | 5219 |

*Source:* Hedman, B., Darrow, K., Wong, E., and Hampson, A. ICF International, 2011. Combined Heat and Power: 2011–2030 Market Assessment. California Energy Commission. CEC-200-2012-002. With permission.

**TABLE 8.22**
Total CHP Potential in 2030 as a Function of Utility Territory

| Utility Region | 50–500 kW (MW) | 500–1 MW (MW) | 1–5 MW (MW) | 5–20 MW (MW) | >20 MW (MW) | Total (MW) |
|---|---|---|---|---|---|---|
| LADWP | 278 | 228 | 355 | 253 | 473 | 1,588 |
| PG&E | 1,234 | 518 | 1,193 | 943 | 3,203 | 7,090 |
| SCE | 1,227 | 441 | 1,013 | 1,074 | 1,236 | 4,991 |
| SDG&E | 265 | 123 | 251 | 152 | 234 | 1,024 |
| SMUD | 98 | 51 | 105 | 153 | 24 | 432 |
| Other North | 68 | 26 | 68 | 78 | 149 | 390 |
| Other South | 125 | 47 | 114 | 163 | 0 | 449 |
| Total (MW) | 3,295 | 1,434 | 3,099 | 2,815 | 5,320 | 15,964 |

*Source:* Hedman, B., Darrow, K., Wong, E., and Hampson, A. ICF International, 2011. Combined Heat and Power: 2011–2030 Market Assessment. California Energy Commission. CEC-200-2012-002. With permission.

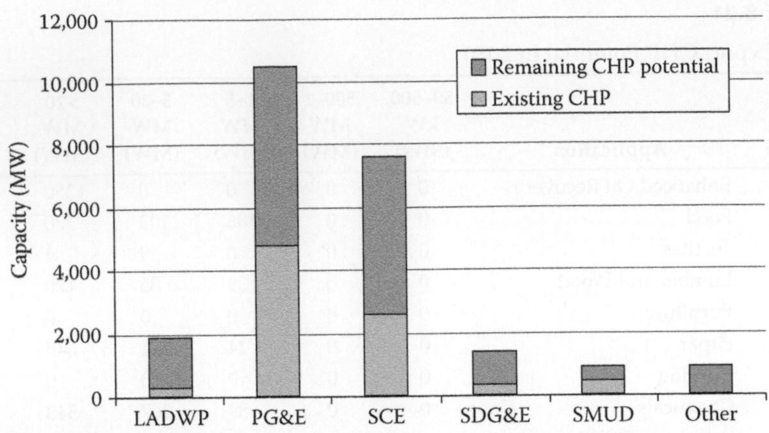


**FIGURE 8.13 (See color insert.)**
The present CHP potential and the total remaining CHP potential as a function of the utility territory. (From ICF International. With permission.)

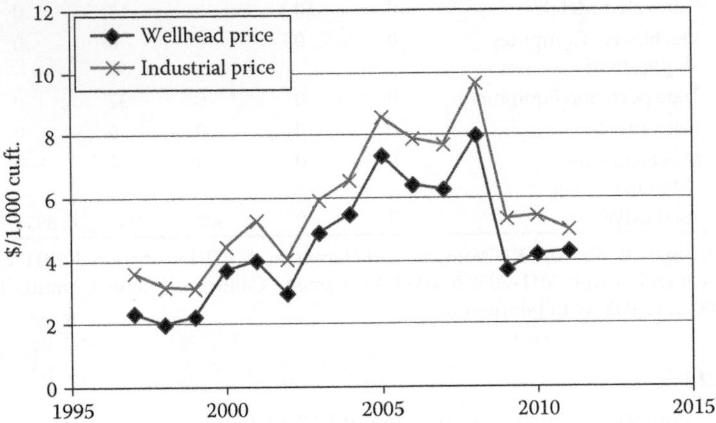


**FIGURE 8.14**
Average U.S. wellhead and industrial natural gas prices between 1997 and 2011. (From Hedman, B., Darrow, K., Wong, E., and Hampson, A. ICF International, 2011. Combined Heat and Power: 2011–2030 Market Assessment. California Energy Commission. CEC-200-2012-002. With permission.)

## 8.6 Natural Gas Pricing

The interdependence between the electric retail and natural gas prices plays a key role for the competitiveness of CHP. This section synthesizes the evolution of gas and electricity prices over a period of 20 years assumed for the present CHP market analysis. It also provides a comparison of the 2011 price assumptions to the 2009 assumptions.

### 8.6.1 Natural Gas Prices

Natural gas prices are decisively influenced by the cost of gas at the wellhead and the cost of transportation to the consumer. The natural gas market today is different compared to that of a few years ago. On one hand, the volatility began in 2000 lasted until 2008, and on the other hand, prices have declined, as shown in Figure 8.14.

The lower prices that followed the 2008 spike can be explained by a pronounced change in the resource outlook of natural gas supply and a less pronounced reduction in demand caused by the global economical recession. Although the long-term demand outlook for natural gas is increasing, this happens at a slow rate. These increases seem to occur mainly in the field of electricity generation. The most important factor expected to keep natural gas prices at a reasonable level in the future is increased production from unconventional sources. In the United States, one of the most efficient unconventional sources is shale gas.

Beginning in 2005, shale gas production increased by approximately 50% per year. This contributed to the doubling of the North American natural gas resource, and resulted in a price of under $5/million BTU. Given the current rates of production and consumption, North American gas resources seem to have the potential to last for another 150 years.

Earlier long-term forecasts, before the increase in economic production of shale gas became evident, were based on a much lower resource base. In addition, marginal supplies in later years were expected to come from much more expensive liquefied natural gas (LNG). Today, valuable natural gas market forecasts (from EIA, the California Energy Commission, and ICF) predict lower volatility and much lower gas prices due to the contribution of, among others, shale gas, which caused an increase in reserves. This contributes further to the removal of LNG as the long-term marginal source of supply, and represents an interesting opportunity for MTs, which over a longer period of time can benefit from this situation.

### 8.6.2 CHP Performance and Cost

Generally, CHP systems use different types of fuel to generate useful heat and electrical energy for the consumer. There is a variety of technologies and products capable of carrying out this function. Although these technologies differ significantly from the configuration and operation points of view, the economic value of CHP depends on key factors that are the same for all CHP technologies:

- Fuel required to generate electricity—usually measured as BTU/kWh. This is the same measurement unit as that for measuring and pricing natural gas. Manufacturers often express engine heat rates in terms of lower heating value (LHV) which precludes the heat of vaporization of the moisture content of the exhaust. That is why the

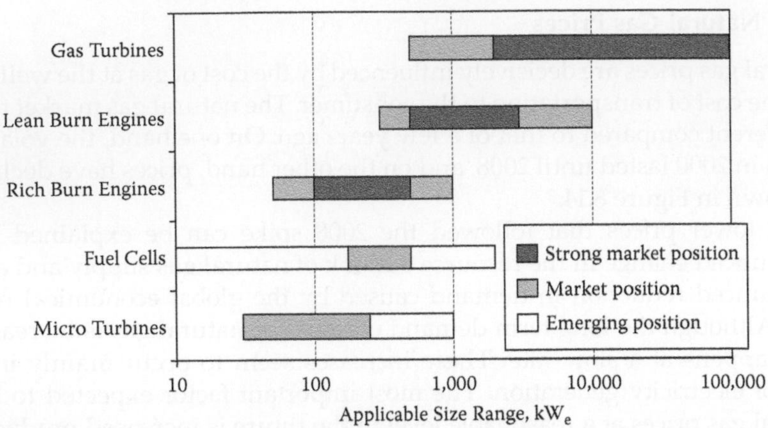


**FIGURE 8.15**
CHP technologies and their competitive market shares. (For more information, please refer to Clean Distributed Generation Performance and Cost Analysis, DE Solutions for ORNL. April 2004.) (From Hedman, B., Darrow, K., Wong, E., and Hampson, A. ICF International, 2011. Combined Heat and Power: 2011–2030 Market Assessment. California Energy Commission. CEC-200-2012-002. With permission.)

manufacturer heat rate and efficiency quotes for natural gas fueled equipments are approximately 10–11% higher.

- Useful thermal energy produced per unit of electrical energy output (again measured as BTU/kWh).
- Installed capital cost of the system, on a unit basis expressed in USD/kWh. Here, an important category of capital costs includes emissions treatment equipment costs, which have to be paid in order to bring some CHP systems into compliance with California emissions requirements.
- Amortization period and the economic life of the equipment.
- Costs not related to either maintenance or operation, expressed again in USD/kWh. Also included are the amortization of overhaul costs which can be required after a number of years of exploitation, as well as the annual costs.
- Emissions treatment capital and operating costs and the pollutant emissions expressed in lb/MWh.

Figure 8.15 presents the different types of CHP technologies together with their competitive market range.

CHP technologies that compete in the California market are as follows:

- Classic gas turbines, which are similar to jet engines and produce high-quality electrical and thermal power for industrial and large commercial consumers. Their capacity varies from a few hundred kilowatts to above 20 MW, but the most economic are in sizes of 5 MW and larger. In large applications of 20 MW and above, they are used almost exclusively for systems using a gaseous fuel. The gas turbines operating under the California environmental regulations must use "alter-treatment" of the exhaust in the form of selective catalytic reduction (SCR).

- The reciprocating engine, which is similar to the type of engine used in most automobiles. Reciprocating engines are available in various sizes, ranging from a few kW to several MW. In Figure 8.15, the reciprocating engines are divided into rich burn and lean burn.
- Rich burn engines are typically offered in smaller sizes and are used for commercial CHP systems of approximately 100 kW. From a historical point of view, rich burn engines have been used in California also as 10 kW units. At present, rich burn engines are sold with integrated emissions control systems, usually a three-way catalyst and an engine control module. The thermal energy generated by these engines is usually available as hot water.
- Lean burn engines operate with excess air to limit nitrogen oxide ($NO_x$) formation and typically are offered in larger sizes. These systems are viable for capacities in the range of 800–5000 kW. Larger engines are available as well. Despite the fact that this technology reduces the emissions of $NO_x$ and of other pollutants, additional "after-treatment" is required to comply with the stringent California emissions requirements. In this case, the thermal energy is available either as hot water or as steam recovery.
- Fuel cells represent a clean class of technology that produces electricity through electrochemical reactions rather than by combustion. Used in the space program or in DG, they are available in many types, according to the different types of chemical construction of their electrolyte (for example, phosphoric acid, molten carbonate, solid oxide, and solid polymer electrolyte). The most common in California are phosphoric acid and molten carbonate cells. Fuel cells are the most expensive source for a CHP application, although it has been promised that improvements will, in the near future, diminish these costs.
- MTs; these have already been described.

In this study, a representative sample of classical and emerging (like the natural gas MTs) CHP systems has been selected to demonstrate the performance and cost characteristics in CHP applications. These systems vary in capacity from approximately 100 to 40,000 kW. The study includes gas-fired reciprocating engines, gas turbines, MTs, and fuel cells. Only the comparison between MTs and reciprocating engines is presented here.

Coming back to the general study, the appropriate technologies were allowed to compete for market share in the penetration model. In the smaller market sizes, as in many of the actual situations, reciprocating engines competed with fuel cells and with MTs. For larger sizes varying from 1 to 20 MW, reciprocating engines competed with gas turbines.

The cost and performance forecasts for the CHP systems, presented below, were based on the work undertaken for the EPA.* Through contacts with manufacturers and developers active in the California market, these estimates were updated for this study. Technology characteristics are presented as 5-year averages over the next 20 years. The costs corresponding to the interval 2010–2015 represent the currently available cost and performance. The year estimates are based on the assumption that continuous improvements in performance and costs will be brought to the CHP applications. The economic features of each of these technologies are presented in the following sections.

### 8.6.3 Emissions Requirements

California has strict emissions standards for CHP equipment. In 2007, the California Air Resources Board imposed the base pollutant emissions standards for fossil fueled DG, as shown in Table 8.23. These standards will apply as well to waste-fueled DG and to biomass, after January 1, 2013. The DG operating as CHP application is allowed to take credit for thermal energy used at the rate of 3.4 millions BTU/MWh—in other words the thermal energy is valued on the same output basis as the electric energy output. In addition, the heat recovery equipment must be integrated to the system and the overall system efficiency must be at least 60% or greater.

All technologies included in this report are capable of meeting this standard. It is clear that the fuel cells comply easily with the standard without after-treatment. As mentioned, gas turbines, reciprocating engines, and

**TABLE 8.23**

ARB Fossil Fuel Emissions Standards from 2007

| Pollutant | Emissions Standard, lb/MWh |
|---|---|
| $NO_x$ | 0.07 |
| CO | 0.10 |
| VOCs | 0.02 |

*Source:* Hedman, B., Darrow, K., Wong, E., and Hampson, A. ICF International, 2011. Combined Heat and Power: 2011–2030 Market Assessment. California Energy Commission. CEC-200-2012-002. With permission.

* For more information on this, please refer to CHP Technology Characterization, EPA CHP Partnership Program, December 2007.

MTs require emissions control systems to clean up the exhaust. Also suitable for the automotive industry, the rich burn engines use a three-way catalyst, which operates much like the catalytic converter in an automobile.

MTs are capable of complying with the standard, due to the advances in Low $NO_x$ combustion that have been recently achieved. Gas turbines and lean burn engines cannot meet the standards using low $NO_x$ combustion alone. Thus, they must use a combination of exhaust gas after-treatment and low $NO_x$ combustion. In this case, the system used is selective catalytic reduction, a process by which the exhaust is treated with ammonia, which reduces the $NO_x$ in the exhaust to nitrogen gas and water vapor. These SCR systems can add up to $300/kW to the cost of the CHP system as well as adding additional operation and maintenance costs.

Some of the most important CHP technologies are represented by reciprocating engines, fuel cells, and natural gas MTs. Among these, the last two are emerging technologies. Reciprocating engines and MTs compete with each other, as seen in Chapter 7. In the following text, these two technologies will be thoroughly investigated from the point of view of the performances and costs.

We begin with the reciprocating engine. Reciprocating engine performance and cost forecasts are shown in Tables 8.24 and 8.25. The tables show the performance and the economic factors that play a key role for the technologies used in the study. The net power cost is determined using the natural gas price forecast described in the section devoted to pricing as well as the existing federal income tax credit for CHP and the California SGIP incentive. Hence, the net power cost is equal to the unit cost of power from the CHP system after the value of the thermal energy has been subtracted. Thermal energy is calculated considering that a boiler operates at 80% efficiency and that between 80 and 100% of the available thermal energy is actually used. Load factors of 80% are estimated for small systems, and load factors of 90% are estimated for large systems. The net capital cost factor is determined based on the economic life of the equipment and a 10% cost of capital. The construction costs vary throughout the state; it has been calculated that the average cost is 6.2% higher than the national average costs. Another assumption of this report is that the real capital costs for smaller reciprocating engines are supposed to diminish by 20% over the next 20 years. The real capital costs for larger reciprocating engine CHP systems are assumed to decline over the next 20 years by 10%. These abatements are expected to be the result of a more competitive market for installation and system design of the technology improvement.

Figure 8.16 compares the net power costs for the reciprocating engine CHP systems given a 20-year market forecast period. As can be seen, net power costs initially decrease and then increase due to the fact that the California SGIP and federal income tax credit (ITC) end while the natural gas prices rise.

The cost and performance forecasts of the natural gas MTs are shown in Table 8.26 and Figure 8.17. MTs are usually suitable in smaller

**TABLE 8.24**

Small Reciprocating Engine Performance and Cost

| CHP System | Characteristics | 2010–2015 | 2016–2020 | 2021–2025 | 2026–2030 |
|---|---|---|---|---|---|
| 100 kW—Rich burn with three-way catalyst | U.S. Average Installed Cost, $/kW | $2,750 | $2,475 | $2,200 | $2,200 |
| | CA Installed Cost, $/kW | $2,921 | $2,629 | $2,337 | $2,337 |
| | After-treatment Cost, $/kW | $0 | $0 | $0 | $0 |
| | Federal Tax Credit, $/kW | $292 | $263 | $0 | $0 |
| | Present Value SGIP, $/kW | $440 | $440 | $0 | $0 |
| | Net Capital Cost, $/kW | $2,190 | $1,927 | $2,337 | $2,337 |
| | O&M, $/kWh | $0.0220 | $0.0200 | $0.0183 | $0.0183 |
| | Heat Rate, Btu/kWh | 12,637 | 11,488 | 10,531 | 10,531 |
| | Useful Thermal, Btu/kWh | 6,700 | 6,091 | 5,583 | 5,583 |
| | CHP Gas Cost, $/MMBtu | $5.44 | $5.75 | $6.53 | $7.25 |
| | Boiler Fuel Gas Cost, $/MMBtu | $7.40 | $7.71 | $8.49 | $9.21 |
| | Net Power Cost, $/kWh | $0.0822 | $0.0752 | $0.0835 | $0.0871 |
| | Economic Life, years | 15 | 15 | 15 | 15 |
| 800 kW—Lean Burn | U.S. Average Installed Cost, $/kW | $1,900 | $1,710 | $1,520 | $1,520 |
| | CA Installed Cost, $/kW | $2,018 | $1,817 | $1,615 | $1,615 |
| | After-treatment Cost, $/kW | $300 | $240 | $180 | $180 |
| | Federal Tax Credit, $/kW | $232 | $206 | $0 | $0 |
| | Present Value SGIP, $/kW | $440 | $440 | $0 | $0 |
| | Net Capital Cost, $/kW | $1,647 | $1,411 | $1,795 | $1,795 |
| | O&M, $/kWh | $0.0160 | $0.0140 | $0.0120 | $0.0120 |
| | Heat Rate, Btu/kWh | 9,760 | 9,750 | 9,225 | 9,225 |
| | Useful Thermal, Btu/kWh | 4,299 | 4,300 | 3,800 | 3,800 |
| | CHP Gas Cost, $/MMBtu | $5.35 | $5.66 | $6.44 | $7.16 |
| | Boiler Fuel Gas Cost, $/MMBtu | $6.98 | $7.28 | $8.07 | $8.79 |
| | Net Power Cost, $/kWh | $0.0691 | $0.0643 | $0.0744 | $0.0783 |
| | Economic Life, years | 15 | 15 | 15 | 15 |

*Source:* Hedman, B., Darrow, K., Wong, E., and Hampson, A. ICF International, 2011. Combined Heat and Power: 2011–2030 Market Assessment. California Energy Commission. CEC-200-2012-002. With permission.

CHP applications. They seem somewhat more costly to purchase and operate than similarly sized reciprocating engine systems, though the comparison is not entirely appropriate since the equipment has different rated powers of 65 kW (MTs) and 100 kW (reciprocating engines). The heat rate measured in BTU/kWh is greater in the case of the MTs.

**TABLE 8.25**

Large Reciprocating Engine Performance and Cost

| CHP System | Characteristics | 2010–2015 | 2016–2020 | 2021–2030 | 2026–2030 |
|---|---|---|---|---|---|
| 3000 kW—Lean Burn | U.S. Average Installed Cost, $/kW | $1,450 | $1,378 | $1,305 | $1,305 |
| | CA Installed Cost, $/kW | $1,540 | $1,463 | $1,386 | $1,386 |
| | After-treatment Cost, $/kW | $200 | $160 | $120 | $120 |
| | Federal Tax Credit, $/kW | $174 | $162 | $0 | $0 |
| | Present Value SGIP, $/kW | $256 | $256 | $0 | $0 |
| | Net Capital Cost, $/kW | $1,310 | $1,205 | $1,506 | $1,506 |
| | O&M, $/kWh | $0.0160 | $0.0152 | $0.0145 | $0.0145 |
| | Heat Rate, Btu/kWh | 9,800 | 9,400 | 9,000 | 9,000 |
| | Useful Thermal, Btu/kWh | 4,200 | 3,850 | 3,500 | 3,500 |
| | CHP Gas Cost, $/MMBtu | $5.33 | $5.63 | $6.42 | $7.14 |
| | Boiler Fuel Gas Cost, $/MMBtu | $6.54 | $6.85 | $7.64 | $8.35 |
| | Net Power Cost, $/kWh | $0.0627 | $0.0620 | $0.0708 | $0.0748 |
| | Economic Life, years | 20 | 20 | 20 | 20 |
| 5000 kW—Lean Burn | U.S. Average Installed Cost, $/kW | $1,450 | $1,378 | $1,305 | $1,305 |
| | CA Installed Cost, $/kW | $1,540 | $1,463 | $1,386 | $1,386 |
| | After-treatment Cost, $/kW | $150 | $120 | $90 | $80 |
| | Federal Tax Credit, $/kW | $169 | $158 | $0 | $0 |
| | Present Value SGIP, $/kW | $103 | $103 | $0 | $0 |
| | Net Capital Cost, $/kW | $1,419 | $1,322 | $1,476 | $1,466 |
| | O&M, $/kWh | $0.0140 | $0.0133 | $0.0127 | $0.0127 |
| | Heat Rate, Btu/kWh | 8,486 | 8,325 | 7,935 | 7,935 |
| | Useful Thermal, Btu/kWh | 3,073 | 2,950 | 2,700 | 2,700 |
| | CHP Gas Cost, $/MMBtu | $5.13 | $5.44 | $6.22 | $6.94 |
| | Boiler Fuel Gas Cost, $/MMBtu | $6.19 | $6.49 | $7.28 | $8.00 |
| | Net Power Cost, $/kWh | $0.0585 | $0.0579 | $0.0633 | $0.0666 |
| | Economic Life, years | 20 | 20 | 20 | 20 |

*Source:* Hedman, B., Darrow, K., Wong, E., and Hampson, A. ICF International, 2011. Combined Heat and Power: 2011–2030 Market Assessment. California Energy Commission. CEC-200-2012-002. With permission.

The base case results of this study show that under the current policy situation, the CHP program will not comply with the ARB Scoping Plan market penetration target. Thus additional policy measures, represented in the medium and high cases, are needed to raise market penetration up to the Scoping Plan target.

Finally, the combined heat and power saves money for the facilities that adopt it. This is the main motivation that drives consumers to adopt it.

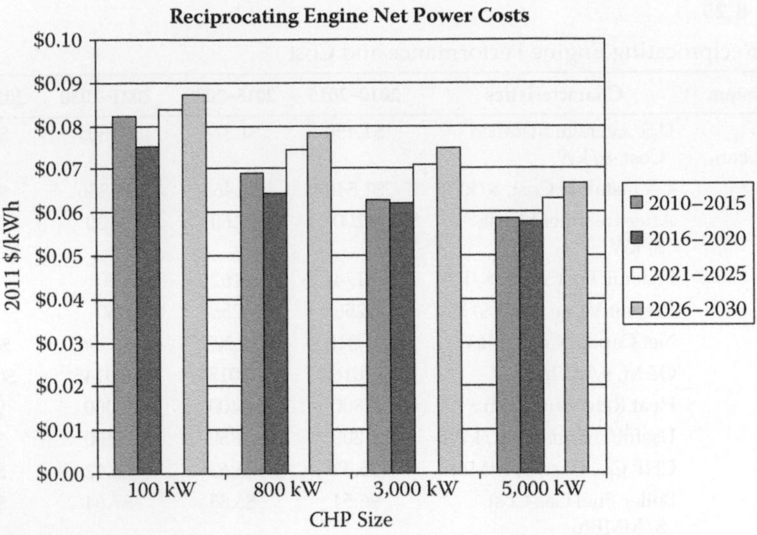


**FIGURE 8.16 (See color insert.)**
Net power costs for reciprocating engines. (From Hedman, B., Darrow, K., Wong, E., and Hampson, A. ICF International, 2011. Combined Heat and Power: 2011–2030 Market Assessment. California Energy Commission. CEC-200-2012-002. With permission.)

**TABLE 8.26**
Microturbine CHP Performance and Cost

| CHP System | Characteristics | 2010–2015 | 2016–2020 | 2021–2030 | 2021–2030 |
|---|---|---|---|---|---|
| 65 kW | U.S. Average Installed Cost, $/kW | $3,100 | $2,790 | $2,480 | $2,480 |
| | CA Installed Cost, $/kW | $3,293 | $2,964 | $2,635 | $2,635 |
| | After-treatment Cost, $/kW | $0 | $0 | $0 | $0 |
| | Federal Tax Credit, $/kW | $329 | $296 | $0 | $0 |
| | Present Value SGIP, $/kW | $440 | $440 | $0 | $0 |
| | Net Capital Cost, $/kW | $2,524 | $2,228 | $2,635 | $2,635 |
| | O&M, $/kWh | $0.0250 | $0.0227 | $0.0208 | $0.0208 |
| | Heat Rate, Btu/kWh | 13,950 | 13,286 | 12,682 | 12,682 |
| | Useful Thermal, Btu/kWh | 5,562 | 5,297 | 5,056 | 5,056 |
| | CHP Gas Cost, $/MMBtu | $5.44 | $5.75 | $6.53 | $7.25 |
| | Boiler Fuel Gas Cost, $/MMBtu | $7.40 | $7.71 | $8.49 | $9.21 |
| | Net Power Cost, $/kWh | $0.1071 | $0.1000 | $0.1101 | $0.1156 |
| | Economic Life, years | 15 | 15 | 15 | 15 |

**TABLE 8.26 (*Continued*)**

Microturbine CHP Performance and Cost

| CHP System | Characteristics | 2010–2015 | 2016–2020 | 2021–2030 | 2021–2030 |
|---|---|---|---|---|---|
| 185 KW | U.S. Average Installed Cost, $/kW | $3,000 | $2,700 | $2,400 | $2,400 |
| | CA Installed Cost, $/kW | $3,187 | $2,868 | $2,550 | $2,550 |
| | After-treatment Cost, $/kW | $0 | $0 | $0 | $0 |
| | Federal Tax Credit, $/kW | $319 | $287 | $0 | $0 |
| | Present Value SGIP, $/kW | $440 | $440 | $0 | $0 |
| | Net Capital Cost, $/kW | $2,429 | $2,142 | $2,550 | $2,550 |
| | O&M, $/kWh | $0.0220 | $0.0200 | $0.0183 | $0.0183 |
| | Heat Rate, Btu/kWh | 12,247 | 11,663 | 11,133 | 11,133 |
| | Useful Thermal, Btu/kWh | 4,265 | 4,062 | 3,877 | 3,877 |
| | CHP Gas Cost, $/MMBtu | $5.44 | $5.75 | $6.53 | $7.25 |
| | Boiler Fuel Gas Cost, $/MMBtu | $7.40 | $7.71 | $8.49 | $9.21 |
| | Net Power Cost, $/kWh | $0.1026 | $0.0959 | $0.1060 | $0.1112 |
| | Economic Life, years | 15 | 15 | 15 | 15 |
| 925 kW | U.S. Average Installed Cost, $/kW | $2,900 | $2,610 | $2,320 | $2,320 |
| | CA Installed Cost, $/kW | $3,081 | $2,773 | $2,465 | $2,465 |
| | After-treatment Cost, $/kW | $0 | $0 | $0 | $0 |
| | Federal Tax Credit, $/kW | $308 | $277 | $0 | $0 |
| | Present Value SGIP, $/kW | $440 | $440 | $0 | $0 |
| | Net Capital Cost, $/kW | $2,333 | $2,056 | $2,465 | $2,465 |
| | O&M, $/kWh | $0.0200 | $0.0182 | $0.0167 | $0.0167 |
| | Heat Rate, Btu/kWh | 12,247 | 11,663 | 11,133 | 11,133 |
| | Useful Thermal, Btu/kWh | 4,265 | 4,062 | 3,877 | 3,877 |
| | CHP Gas Cost, $/MMBtu | $5.33 | $5.63 | $6.42 | $7.14 |
| | Boiler Fuel Gas Cost, $/MMBtu | $6.54 | $6.85 | $7.64 | $8.35 |
| | Net Power Cost, $/kWh | $0.1011 | $0.0946 | $0.1048 | $0.1100 |
| | Economic Life, years | 15 | 15 | 15 | 15 |

*Source:* Hedman, B., Darrow, K., Wong, E., and Hampson, A. ICF International, 2011. Combined Heat and Power: 2011–2030 Market Assessment. California Energy Commission. CEC-200-2012-002. With permission.

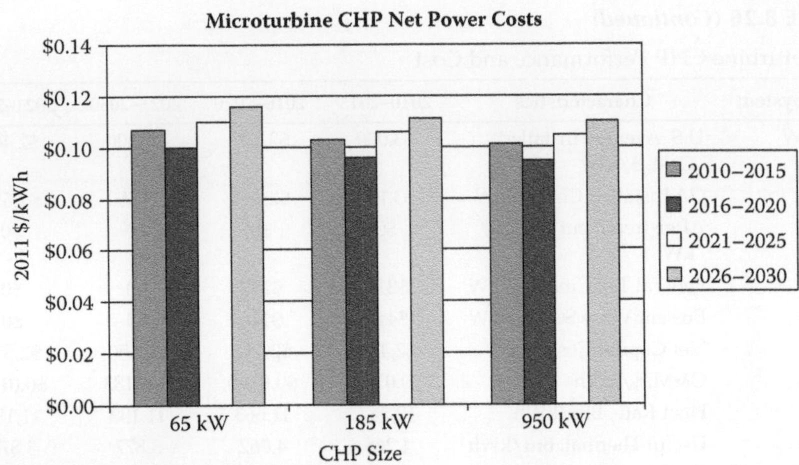


**FIGURE 8.17 (See color insert.)**
MT CHP net power costs. (From Hedman, B., Darrow, K., Wong, E., and Hampson, A. ICF International, 2011. Combined Heat and Power: 2011–2030 Market Assessment. California Energy Commission. CEC-200-2012-002. With permission.)

According to the studies, by 2030, combined heat and power would save customers $740 million a year in energy costs under the base case and $2.9 billion a year under the high case. Other measures that provide social benefits such as transmission and distribution capacity postponing, greenhouse gas emissions reduction, and energy efficiency will increase market penetration for CHP as shown by the market response in the medium and high cases analyzed.

For further information on market potential for CHP applications in California, please refer to [HKW12].

## 8.7 Concluding Remarks

Despite the fact that we have come a long way since the Heron's Aeolipile in the first century AD, which is considered to be the precursor of the modern gas turbine, much remains to be done in this field. This strange looking instrument, which at that time had no practical use, began to find some appreciation only in 1791 when the English inventor John Barber proposed a propulsion method for a horseless carriage, though the technology of that era did not yet permit practical application. John Barber is credited as the first person to have described in great detail the working principle of a gas turbine. Successful construction of a gas turbine came sometime late in 1904, when Frank Stolze, in Berlin, designed a piece of equipment that had little efficiency. It was not until 1937 that the first gas turbine to be used in jet propulsion was patented in England by Sir Frank Whittle.

Over time, many improvements have been brought to Heron's device, but it has not yet reached its full potential. Between Heron and Leonardo da Vinci's smoke mill in 1550, there is a hiatus of approximately 1480 years in which nothing further was developed. In addition, the technology at the time did not permit practical application. At present, therefore, gas turbogenerators are living their second infancy.

# References

[AAH04] A. Hartzheim, Multiple control loop acceleration of turboalternator after reaching self-sustaining speed previous to reaching synchronous speed. U.S. Patent 6,834,226, issued December 21, 2004, available online: http://www.uspto.gov (accessed March 20, 2013).

[AC05] A. Amorim and A.L. Cardoso, Analysis of the connection of a microturbine to a low voltage grid, International Conference on Future Power Systems, Amsterdam, November 18, 2005, pages 5–10.

[AFM69] A.F. McClean et al., Gas turbine control system. U.S. Patent 3,485,042 issued December 23, 1969, available online: http://www.uspto.gov (accessed February 25, 2013).

[AHB76] A. Bell III et al., Automotive gas turbine control. U.S. Patent 3,999,373, issued December 28, 1976, available online: http://www.uspto.gov (accessed March 1, 2013).

[AHm04] A. Hartzheim, Method for ignition and start up of a turbogenerator. U.S. Patent 6,766,647, issued July 27, 2004, available online: http://www.uspto.gov (accessed March 19, 2013).

[ALo70] A. Loft et al., Gas turbine control system. U.S. Patent 3,520,133 issued July 14, 1970, available online: http://www.uspto.gov (accessed February 28, 2013).

[AM69] A.F. McClean, Control system for gas turbine/transmission power train. U.S. Patent 3,486,329 issued December 30, 1969, available online: http://www.uspto.gov (accessed February 25, 2013).

[AMP54] A.M. Prentiss, Fuel and speed control apparatus. U.S. Patent 2,691,268 issued October 12, 1954, available online: http://www.uspto.gov (accessed February 25, 2013).

[APA07] A.B.M. Aguiar, J.O.P. Pinto, C.Q. Andrea, and L.A. Nogueira, Modeling and simulation of natural gas microturbine application for residential complex aiming technical and economical viability analysis, Electrical Power Conference, Canada, October 25–26, 2007, pages 376–381.

[APM74] A.P. McClean et al., Gas turbine control system. U.S. Patent 3,795,104 issued March 5, 1974, available online: http://www.uspto.gov (accessed February 25, 2013).

[ASV68] A. Silver et al., Selectively pressurized foil bearing arrangements. U.S. Patent 3,366,427, issued January 30, 1968, available online: http://www.uspto.gov (accessed March 12, 2013).

[ASV79] A. Silver et al., Foil bearing. U.S. Patent 4,178,046, issued December 11, 1979, available online: http://www.uspto.gov (accessed March 14, 2013).

[AUS-- ] Information on the Austin gas turbine vehicle, available online: http://www.austinmemories.com/page19/page19.html (accessed November 28, 2012).

[AW04] J. Arrillaga and N.R. Watson, *Power System Harmonics*, Wiley, London, 2004.

[BBC-- ] Information on the Rover Jet 1 car, available online: http://news.bbc.co.uk/onthisday/hi/dates/stories/march/8/newsid_2516000/2516271.stm (accessed April 29, 2013).

[BCM08] A.V. Boicea, G. Chicco, and P. Mancarella, Assessing the performance of microturbine clusters, *Proceedings of the VII World Energy System Conference*, Iasi, Romania, June 30–July 2, 2008.

[BCM09] A.V. Boicea, G. Chicco, and P. Mancarella, Optimal operation of a microturbine cluster with partial-load efficiency and emission characterization, Powertech, 2009 IEEE Bucharest, June 28–July 2, 2009, pages 1–8.

[BCM11] A.V. Boicea, G. Chicco, and P. Mancarella, Optimal operation of a 30 kW natural gas microturbine cluster, *Buletinul Stiintific al Universitatii Politehnica Bucuresti, Seria C*, 73(1), 211–222, 2011.

[BDy02] B. Dickey et al., Ultra low emissions gas turbine cycle using variable combustion primary zone airflow control. U.S. Patent 20020104316, issued August 8, 2002, available online: http://www.uspto.gov (accessed March 16, 2013).

[BE01] R. Bosley et al., Turbogenerator power control system. U.S. Patent 20010052704, issued December 20, 2001, available online: http://www.uspto.gov (accessed March 20, 2013).

[BEL02] R.W. Bosley et al., Turbogenerator power control system. U.S. Patent 6,495,929, issued December 17, 2002, available online: http://www.uspto.gov (accessed March 20, 2013).

[BGD95] B.G. Donnelly et al., Centrifugal pump with starting stage. U.S. Patent 5,456,574, issued October 10, 1995, available online: http://www.uspto.gov (accessed March 7, 2013).

[Bol-- ] Information on the Capstone microturbines in Bolivia, available online: http://www.microturbine.com/_docs/CS_CAP394_Bolivian%20Pipeline_lowres.pdf (accessed April 5, 2013).

[Cal-- ] Information on the Elliott (Calnetix) gas microturbine, available online: http://www.tai-cepc.ro/fileadmin/download/pdf/Leafleats%20Capstone%20%26%20Calnetix/CalTA100.pdf (accessed April 29, 2013).

[Car-- ] Information on the Capstone microturbines in Carneys Point, New Jersey, available online: http://www.microturbine.com/_docs/CS_CAP391_SalemCC_lowres.pdf (accessed April 11, 2013).

[CC06] G.W. Chang and Y.C. Chin, Efficient approach to characterising harmonic currents generated by a cluster of three-phase AC/DC converters, *IEE Proceedings on Electric Power Applications*, Volume 153, Issue 5, September 2006, pages 742–749.

[CCM07] A. Canova, G. Chicco, and P. Mancarella, Assessment of the emissions due to cogeneration microturbines under different operation modes, *Proceedings of Powereng 2007*, Setúbal, Portugal, April 2007, pages 684–689.

[CdB-- ] Information on the Capstone microturbines at Cavenago di Brianza, Italy, available online: http://www.microturbine.com/_docs/CS_CAP346_Milan_Italy_lowres.pdf (accessed March 21, 2013).

[CEJ80] C.E. Jahnig et al., Gas turbine power system with fuel injection and combustion catalyst. U.S. Patent 4,197,700, issued April 15, 1980, available online: http://www.uspto.gov (accessed March 17, 2013).

[Cha65] A. Chadwick, Fuel control system for a gas turbine engine. U.S. Patent 3,183,667, issued May 18, 1965, available online: http://www.uspto.gov (accessed February 25, 2013).

[Cio-- ] Information on the Capstone microturbines at La Ciotat, France, available online: http://www.microturbine.com/_docs/CS_CAP408_LaCiotat_lowres.pdf (accessed April 3, 2013).

[CKS-- ] Information on Firebird III developed by General Motors, available online: http://www.conklinsystems.com/firebird/oldnews.php (accessed January 25, 2013).

[CL95] M.T. Chen, J.W. Lue, and H.Y. Chu, Statistical standards and pollution estimation of harmonics in a power system, International Conference on EMPD (Energy Management and Power Delivery), November 21–23, 1995, pages 662–667.

[CMN07] G. Chicco, P. Mancarella, and R. Napoli, Emission assessment of distributed generation in urban areas, *Proceedings of IEEE Power Tech 2007*, Lausanne, Switzerland, July 1–5, 2007, pages 532–537.

[CN94] D.E. Campbell and R. Nagahisa, A foundation for Pareto aggregation, *Journal of Economic Theory*, 64, 277–285, 1994.

[COG-- ] Information on the Capstone microturbines at Cognac, France, available online: http://www.microturbine.com/_docs/CS_CAP396_Revico_lowres.pdf (accessed April 1, 2013).

[CPT-- ] Information on the Capstone and Calnetix (former Elliott) microturbines, available online: http://www.capstoneturbine.com/news/story.asp?id = 545 (accessed March 8, 2013).

[CPTC-- ] Information on the Capstone compressor, available online: http://www.capstoneturbine.com/news/story.asp?id = 119 (accessed March 18, 2013).

[CSK59] C.S. King et al., Fuel supply system for a gas turbine engine power plant. U.S. Patent 2,874,765, issued February 24, 1959, available online: http://www.uspto.gov (accessed March 4, 2013).

[CSO-- ] Cogeneration and On–Site Power Production. 2011. Romanian hospital orders CHP microturbine, available online: http://www.cospp.com/articles/2011/08/romanian-hospotal-orders-chp-microtrubine.html (accessed March 21, 2013).

[CSP-- ] Information on the Capstone microturbines at Cossato Spolina, Italy, available online: http://www.microturbine.com/_docs/CS_CAP385_CossatoSpolinaWWTP_lowres.pdf (accessed March 25, 2013).

[DBr46] D. Bradbury, Turbine apparatus. U.S. Patent 2,404,428, issued July 23, 1946, available online: http://www.uspto.gov (accessed March 4, 2013).

[DGB91] D.G. Bridgnell et al., Stress relief for an annular recuperator. U.S. Patent 5,050,668, issued September 24, 1991, available online: http://www.uspto.gov (accessed March 16, 2013).

[DHn04] D. Hamrin et al., Method for catalytic combustion in a gas-turbine engine, and applications thereof. U.S. Patent 20040148942, issued August 5, 2004, available online: http://www.uspto.gov (accessed March 16, 2013).

[DHW06] D.H. Weissert, Compliant foil fluid film radial bearing. U.S. Patent RE39,190, issued July 18, 2006, available online: http://www.uspto.gov (accessed March 12, 2013).

[DJM68] D.J. Marley, Self-acting foil bearings. U.S. Patent 3,382,014, issued May 7, 1968, available online: http://www.uspto.gov (accessed March 12, 2013).

[DJn00] D. Jensen, Microturbine power generating system including variable-speed gas compressor. U.S. Patent 6,066,898, issued May 23, 2000, available online: http://www.uspto.gov (accessed March 17, 2013).

[DPA02] K. Deb, A. Pratap, S. Agarwal, and T. Meyarivan, A fast and elitist multiobjective genetic algorithm: NSGA-II, *IEEE Transactions on Evolutionary Computation*, 6(2), April 2002, 182–197.

[DWD04] D.W. Dewis, Recuperator configuration. U.S. Patent 6,832,470, issued December 21, 2004, available online: http://www.uspto.gov (accessed March 21, 2013).

[ECE00] E.C. Edelman, Gas turbine engine fixed speed light-off method. U.S. Patent 6,062,016, issued May 16, 2000, available online: http://www.uspto.gov (accessed March 8, 2013).

[EDU-- ] EDUCOGEN, The European Educational Tool on Cogeneration, December 2001, available online: www.cogen.org/projects/educogen.htm (accessed March 9, 2013).

[EGS69] E.G. Smith et al., Automatic starting and protection system for a gas turbine. U.S. Patent 3,470,691, issued October 7, 1969, available online: http://www.uspto.gov (accessed March 20, 2013).

[EJB69] E.J. Bevers et al., Turbine governor. U.S. Patent 3,439,496, issued April 22, 1969, available online: http://www.uspto.gov (accessed March 3, 2013).

[EPA-- ] U.S. Environmental Protection Agency (EPA), available online: www.epa.org (accessed March 19, 2013).

[ESH74] E.S. Harrisson et al., Gas turbine start-up fuel control system. U.S. Patent 3,844,112, issued October 29, 1974, available online: http://www.uspto.gov (accessed March 10, 2013).

[FDA04] O. Fethi, L.-A. Dessaint, and K. Al-Haddad, Modeling and simulation of the electric part of a grid connected microturbine, Power Engineering Society General Meeting, Vol. 2, June 10, 2004, pages 2212–2219.

[FHV59] F.H. Van Nest et al., Control of heat supply to heat recovery boiler of regenerative cycle gas turbine powerplant. U.S. Patent 2,914,917, issued December 1, 1959, available online: http://www.uspto.gov (accessed March 22, 2013).

[FMC-- ] Ford Motor Company, available online: http://media.ford.com/article_display.cfm?article_id=14119 (accessed January 28, 2013).

[FRR62] F.R. Rogers et al., Fuel control apparatus for a combustion engine. U.S. Patent 3,050,941, issued August 28, 1962, available online: http://www.uspto.gov (accessed March 3, 2013).

[GCa76] G. Caire et al., Device for controlling gas turbine engines. U.S. Patent 3,987,620, issued October 26, 1976, available online: http://www.uspto.gov (accessed February 27, 2013).

[GDL06] X. Guillaud, P. Degobert, C. Larose, and A. Vallee, Methodological approach for real-time power system simulation application to the connection of a micro turbine generator in a distribution network, IEEE International Symposium on Industrial Electronics, Vol. 3, July 9–13, 2006, pages 2597–2602.

[GHh89] G. Heidrich, Diaphragm disk assembly for torque transmitting joint and process for its fabrication. U.S. Patent 4,802,882, issued February 7, 1989, available online: http://www.uspto.gov (accessed March 11, 2013).

[GMH-- ] Information on Firebird II developed by General Motors, available online: http://www.gmheritagecenter.com/docs/gm-heritage-archive/historical-brochures/1956-firebird-II/1956_Firebird_II_Brochure.pdf (accessed January 29, 2013).

[Go89] D.E. Goldberg, *Genetic Algorithms in Search, Optimization and Machine Learning,* Addison-Wesley, Reading, MA, 1989.

[GoM-- ] Information on the Capstone microturbines in the Gulf of Mexico, Mexico, available online: http://www.microturbine.com/_docs/CS_CAP342_Pemex,Mexico.pdf (accessed April 6, 2013).

[Gov-- ] Information on the Capstone microturbines in Southern USA, available online: http://www.microturbine.com/_docs/CA%20CAP383_US%20Govt%20Office.pdf (accessed April 9, 2013).

[GP06] D.N. Gaonkar and R.N. Patel, Dynamic model of microturbine generation system for grid connected/islanding operation, IEEE International Conference on Industrial Technology (ICIT 2006), Mumbai, December 15–17, 2006, pages 305–310.

[GuDL06] X. Guillaud, P. Degobert, D. Loriol, and E. Mogos, Real-time simulation of a micro-turbine integrated in a distribution network, International Symposium on Power Electronics, Electrical Drives, Automation and Motion, May 23–26, 2006, pages 487–491.

[H97] J.H. Horlock, *Cogeneration—Combined Heat and Power,* Krieger, Malabar, FL, 1997.

[Ha03] S. Hamilton, *Microturbine Generator Handbook,* PennWell, Tulsa, OK, 2003.

[HEF81] H.E. Fogg et al., Flexible coupling. U.S. Patent 4,265,099, issued May 5, 1981, available online: http://www.uspto.gov (accessed March 11, 2013).

[HHa02] H. Hirata et al., Turbine system having a reformer and method thereof. U.S. Patent 6,338,239, issued January 15, 2002, available online: http://www.uspto.gov (accessed March 19, 2013).

[HKW12] Hedman, B., Darrow, K., Wong, E., and Hampson, A. ICF International, 2011. Combined Heat and Power: 2011–2030 Market Assessment. California Energy Commission. CEC-200-2012-002.

[Hol74] B. W. Holleboom et al., Electronic start control circuit for gas turbine engine. U.S. Patent 3,793,826, issued February 26, 1974, available online: http://www.uspto.gov (accessed March 10, 2013).

[HWB94] H.W. Brockner et al., Automotive fuel pump housing with rotary pumping element. U.S. Patent 5,310,308, issued May 10, 1994, available online: http://www.uspto.gov (accessed March 7, 2013).

[HWV59] H.W. Van Gerpen, Hydraulic apparatus. U.S. Patent 2,892,311, issued June 30, 1959, available online: http://www.uspto.gov (accessed March 5, 2013).

[IEC02] Electromagnetic Compatibility (EMC) Standard, Part 4–7: Testing and measurement techniques—General guide on harmonics and interharmonics measurements and instrumentation, for power supply systems and equipment connected thereto, International Electrotechnical Commission, 2002.

[IHM03] M. Izhar, C.M. Hadzer, S. Masri, and S. Idris, A study of the fundamental principles to power system harmonic, Power Engineering Conference, December 15–16, 2003, pages 225–232.

[ILL73] I.L. Lehman, Flexible shaft stabilizer. U.S. Patent 3,779,451, issued December 18, 1973, available online: http://www.uspto.gov (accessed March 11, 2013).

[JAK73] J.A. Karol, Actuating device for a gas turbine engine fuel control. U.S. Patent 3,733,815, issued May 22, 1973, available online: http://www.uspto.gov (accessed March 3, 2013).

[JBI04] J.B. Ingram, Microturbine with auxiliary air tubes for NOx emission reduction. U.S. Patent 6,729,141, issued May 4, 2004, available online: http://www.uspto.gov (accessed March 23, 2013).

[JDe74] J. Delahaye et al., Gas turbine prime mover. U.S. Patent 3,798,898, issued March 26, 1974, available online: http://www.uspto.gov (accessed March 16, 2013).

[JLi00] J. Lipinski et al., Low NOx conditioner system for a microturbine power generating system. U.S. Patent 6,125,625, issued October 3, 2000, available online: http://www.uspto.gov (accessed March 23, 2013).

[JMT01] J.M. Teets et al., Electricity generating system having an annular combustor. U.S. Patent 6,314,717, issued November 13, 2001, available online: http://www.uspto.gov (accessed March 21, 2013).

[JNa74] J. Nicita et al., Gas turbine engine and counterflow heat exchanger with outer air passageway. U.S. Patent 3,831,374, issued August 27, 1974, available online: http://www.uspto.gov (accessed March 16, 2013).

[JNS92] J.N. Stocco, Flexible coupling including a flexible diaphragm element contoured with its thinnest thickness near the center thereof. U.S. Patent 5,158,504, issued October 27, 1992, available online: http://www.uspto.gov (accessed March 12, 2013).

[JTM74] J.T. Moehring et al., Light-off transient control for an augmented gas turbine engine. U.S. Patent 3,834,160, issued September 10, 1974, available online: http://www.uspto.gov (accessed March 10, 2013).

[KAB-- ] Information on the Capstone microturbines at Kupferzell, Germany, available online: http://www.microturbine.com/_docs/CS_CAP395_Kupferzell%20Biogas%20Plant_lowres.pdf (accessed April 1, 2013).

[KCS06] A. Konak, D.W. Coir, and A.E. Smith, Multi-objective optimization using genetic algorithms: A tutorial, *Reliability Engineering and System Safety*, 91, 992–1007, 2006.

[Kne99] T.A. Kneschke, Distortion and power factor of nonlinear loads, Railroad Conference, April 13–15, 1999, pages 47–54.

[Ko04] B.F. Kolanowski, *Guide to Microturbines*, Marcel Dekker, New York, 2004.

[KRC95] K.R. Carr et al., Engine starting system utilizing multiple controlled acceleration rates. U.S. Patent 5,430,362, issued July 4, 1995, available online: http://www.uspto.gov (accessed March 21, 2013).

[LCH06] Y.D. Lee, C.S. Chen, C.T. Hsu, and H.S. Cheng, Harmonic analysis for the distribution system with dispersed generation systems, International Conference on Power System Technology, October 22–26, 2006, pages 1–6.

[Lo96] G. Lozza, Gas turbines and combined cycles (in Italian), Progetto Leonardo, Bologna, Italy, 1996.

[M-- ] Information about the displacement power factor, available online: http://www.markempson.com/motor-control/power-factor/displacement-power-factor (accessed May 2, 2013).

[Man-- ] Information on the Capstone microturbines in Manhattan, New York, available online: http://www.microturbine.com/_docs/CS_CAP380_Ave%20of%20Americas.pdf (accessed April 9, 2013).

[MAT-- ] Information on FIAT Turbina, available online: http://www.museo-auto.it/website/en/collezione/34-fiat/99-fiat-mod-turbina (accessed January 29, 2013).

[MBC93] M.B. Colket III et al., Method and system for combusting hydrocarbon fuels with low pollutant emissions by controllably extracting heat from the catalytic oxidation stage. U.S. Patent 5,235,804, issued August 17, 1993, available online: http://www.uspto.gov (accessed March 17, 2013).

[MG98] A. Mansoor and W.M. Grady, Analysis of compensation factors influencing the net harmonic current produced by single-phase nonlinear loads, Eighth International Conference on Harmonics and Quality of Power, Vol. 2, October 14–16, 1998, pages 883–889.

[MHu95] M. Huebel et al., Peripheral pump, particularly for feeding fuel to an internal combustion engine from a fuel tank of a motor vehicle. U.S. Patent 5,468,119, issued November 21, 1995, available online: http://www.uspto.gov (accessed March 8, 2013).

[MJK92] M.J. Khinkis, Ultra-low pollutant emission combustion method and apparatus. U.S. Patent 5,158,445, issued October 27, 1992, available online: http://www.uspto.gov (accessed March 23, 2013).

[Moh-- ] Information on the Capstone microturbines in the Village of Mohsogollokh, Russia, available online: http://www.microturbine.com/_docs/CS_CAP405_Yakutcement_lores.pdf (accessed April 6, 2013).

[Moo02] M.J. Moore, *Micro–turbine Generators,* Professional Engineering, Bury–St. Edmonds, 2002.

[MTT-- ] Information on Chrysler turbine engine car, available online: http://www.motortrend.com/classic/features/c12_0603_1964_chrysler_turbine_car/viewall.html (accessed February 25, 2013).

[NG97] T. Nguyen, Parametric harmonic analysis [of power systems], *IEE Proceedings—Generation, Transmission and Distribution,* Vol. 144, Issue. 1, January 1997, pages 21–25.

[NL03] D.K. Nichols and K.P. Loving, Assessment of microturbine generators, Power Engineering Society General Meeting, Vol. 4, July 13–17, 2003, pages 2314–2315.

[ORNL-- ] R. Staunton, B. Ozpineci, T.J. Theiss, and L.M. Tolbert, Review of the state of the art in power electronics suitable for 10 kW military power systems, available online: http://www.ornl.gov/~webworks/cppr/y2001/rpt/118205.pdf (accessed May 2, 2013).

[PBK62] P.B. Kahn, Fuel control governors. U.S. Patent 3,035,592, issued May 22, 1962, available online: http://www.uspto.gov (accessed March 5, 2013).

[PBV02] P.B. Vessa, Double diaphragm compound shaft. U.S. Patent 20020079760, issued June 27, 2002, available online: http://www.uspto.gov (accessed March 10, 2013).

[PGL64] P.G. La Haye, Centrifugal compressing of low molecular weight gases. U.S. Patent 3,161,020, issued December 15, 1964, available online: http://www.uspto.gov (accessed March 18, 2013).

[PZR01] A.Y. Petrov, A. Zaltash, T.D. Rizy, and S.D. Labinov, Environmental aspects of operation of a gas-fired microturbine-based CHP system, Oak Ridge National Laboratory report, 2001, available online: http://www.ornl.gov/~webworks/cppr/y2001/pres/115331.pdf (accessed May 2, 2013).

[Q4C-- ] Information on the Capstone microturbines at Q4C Oil Platform, available online: http://www.microturbine.com/_docs/CS_CAP403_Q4C%20Platform_lowres1.pdf (accessed May 2, 2013).

[QZ06] Q. Liang and Z. Hui, An improved modulation of the selective harmonic elimination controlling, International Conference on Power System Technology, October 22–26, 2006, pages 1–5.

[Rav-- ] Information on the Capstone microturbines in Ravenna, USA, available online: http://www.microturbine.com/_docs/CA%20CAP382_den%20Dulk%20Dairy_lowres.pdf (accessed April 8, 2013).

[RBo61] R. Bodemuller et al., Gas turbine acceleration control. U.S. Patent 2,971,338 issued February 14, 1961, available online: http://www.uspto.gov (accessed February 28, 2013).

[RFS00] R.F. Stokes et al., Gas turbine starter assist torque control system. U.S. Patent 6,035,626, issued March 14, 2000, available online: http://www.uspto.gov (accessed March 20, 2013).

[RGk72] R. Gottschalk et al., Self-pressurizing bearings with resilient elements. U.S. Patent 3,635,534, issued January 18, 1972, available online: http://www.uspto.gov (accessed March 14, 2013).

[RH54] R. Hill, System for controlling gas temperatures. U.S. Patent 2,697,328 issued December 21, 1954, available online: http://www.uspto.gov (accessed February 25, 2013).

[RLa48] R. Lapsley, Fluid pump and control therefore. U.S. Patent 2,433,954, issued January 6, 1948, available online: http://www.uspto.gov (accessed March 4, 2013).

[RNM06] M.A. Rendon, M.A.R. Nascimento, and P.P.C. Mendes, Load current control model for a gas micro-turbine in isolated operation, Power Systems Conference and Exposition, October 29–November 1, 2006, pages 1139–1149.

[RNP63] R.N. Penny, Gas turbine engine fuel system. U.S. Patent 3,085,619, issued April 16, 1963, available online: http://www.uspto.gov (accessed March 4, 2013).

[RNP71] R.N. Penny, Rotary fuel pump. U.S. Patent 3,594,100, issued July 2, 1971, available online: http://www.uspto.gov (accessed March 4, 2013).

[Rot-- ] Information on the Capstone microturbines at Rotterdam, Netherlands, available online: http://www.microturbine.com/_docs/CS_CAP406_Argonon_lowres.pdf (accessed April 3, 2013).

[RPL81] R.P. Lohmann et al., Fuel injection system for low emission burners. U.S. Patent 4,265,615, issued May 5, 1981, available online: http://www.uspto.gov (accessed March 23, 2013).

[RR-- ] Information on the Rolls Royce gas turbine engines, available online: http://www.rolls-royce.com/marine/products/diesels_gas_turbines/gas_turbines/ (accessed February 26, 2013).

[RUm77] R. Uram, Accurate, stable and highly responsive gas turbine startup speed control with fixed time acceleration especially useful in combined cycle electric power plants. U.S. Patent 4,010,605, issued March 8, 1977, available online: http://www.uspto.gov (accessed March 20, 2013).

[RWB99] R.W. Bosley, Helical flow compressor/turbine permanent magnet motor/generator. U.S. Patent 5,899,673, issued May 4, 1999, available online: http://www.uspto.gov (accessed March 7, 2013).

[RWK83] R.W. Kiscaden et al., System and method for accelerating and sequencing industrial gas turbine apparatus and gas turbine electric power plants preferably with a digital computer control system. U.S. Patent 4,380,146, issued April 19, 1983, available online: http://www.uspto.gov (accessed March 20, 2013).

[SCC08] A.K. Saha, S. Chowdhury, S.P. Chowdhury, and P.A. Crossley, Modelling and simulation of microturbine in islanded and grid-connected mode as distributed energy resource, Power and Energy Society General Meeting—Conversion and Delivery of Electrical Energy in the 21st Century, July 20–24, 2008, pages 1–7.

[SD07] P.K. Shukla and K. Deb, On finding multiple Pareto-optimal solutions using classical and evolutionary generating methods, *European Journal of Operational Research,* 181, 1630–1652, 2007.

[SFo80] S. Forster et al., Vehicular propulsion gas turbine motor. U.S. Patent 4,213,297, issued July 22, 1980, available online: http://www.uspto.gov (accessed March 22, 2013).

[SFr80] S. Forster et al., Vehicular gas turbine installation with ceramic recuperative heat exchanger elements arranged in rings around compressor, gas turbine and combustion chamber. U.S. Patent 4,180,973, issued January 1, 1980, available online: http://www.uspto.gov (accessed March 22, 2013).

[She-- ] Information on the Capstone microturbines in Sheboygan, Wisconsin, USA, available online: http://www.microturbine.com/_docs/CS_CAP381_Sheboygan_lowres.pdf (accessed April 10, 2013).

[SHm53] S. Holm, Combined regenerator and precooler for gas turbine cycles. U.S. Patent 2,650,073, issued August 25, 1953, available online: http://www.uspto.gov (accessed March 16, 2013).

[Sim-- ] Information on the Capstone microturbines in Simi Valley, USA, available online: http://www.microturbine.com/_docs/CS_CAP386_RRPL_lowres.pdf (accessed April 8, 2013).

[SJH-- ] Information on the Capstone microturbines at St. Jospeh Hospital in Prüm, Germany, available online: http://www.microturbine.com/_docs/CS_CAP398_StJoseph_lowres.pdf (accessed April 1, 2013).

[Smi02] C.W. Smith Jr., Power systems and harmonic factors, *IEEE Potentials,* 20(5), 10–12, December 2001–January 2002.

[SMP-- ] Information on the Capstone microturbines at St. Martin in Passeier, Italy, available online: http://www.microturbine.com/_docs/CS_CAP400_Quellenhof%20Resort_lowres.pdf (accessed March 30, 2013).

[So07] C. Soares, Microturbines—Applications for distributed energy systems, Butterworth–Heinemann, Burlington, MA, 2007.

[STP-- ] Information on the Capstone microturbines in the St. Petersburg region, Russia, available online: http://www.microturbine.com/_docs/CS_CAP388_RussianSkiResort.pdf (accessed April 6, 2013).

[SVv05] S. Voinov, Gas compression system and method for microturbine application. U.S. Patent 6,892,542, issued May 17, 2005, available online: http://www.uspto.gov (accessed March 17, 2013).

[TAo83] T. Abo et al., Fuel control system for gas turbine engine. U.S. Patent 4,378,673, issued April 5, 1983, available online: http://www.uspto.gov (accessed March 20, 2013).

[TC08] C. Tautiva and A. Cadena, Optimal placement of distributed generation on distribution networks, Transmission and Distribution Conference and Exposition: Latin America, August 13–15, 2008, pages 1–5.

[TFa09] T. Furuya et al., Electric power supply system. U.S. Patent 7,514,813, issued April 7, 2009, available online: http://www.uspto.gov (accessed March 22, 2013).

[THM64] T.H. Homes, Sequencing system. U.S. Patent 3,135,088, issued June 2, 1964, available online: http://www.uspto.gov (accessed February 27, 2013).

[TKN98] M. Tsukamoto, I. Kouda, and Y. Nasuda, Advanced method to identify harmonics characteristic between utility grid and harmonic current sources, Eighth International Conference on Harmonics and Quality of Power, October 14–16, 1998, pages 419–425.

[TON00] M. Tsukamoto, S. Ogawa, and Y. Natsuda, Advanced technology to identify harmonics characteristics and results of measuring, Ninth International Conference on Harmonics and Quality of Power, Vol. 1, October 1–4, 2000, pages 341–346.

[Tro-- ] Information on the Capstone microturbines mounted on the Trolza ECObus–5250, Russia, available online: http://www.microturbine.com/_docs/CS_CAP407_Trolza_lowres.pdf (accessed April 8, 2013).

[Ukh-- ] Information on the Capstone microturbines in the City of Ukhta, Russia, available online: http://www.microturbine.com/_docs/CS_CAP404_Yarmarka_lowres.pdf (accessed April 6, 2013).

[VB09] S. Vlahinic, D. Brnobic, and N. Stojkovic, Indices for harmonic distortion monitoring of power distribution systems, *IEEE Transactions on Instrumentation and Measurement*, 58(5), 1771–1777, May 2009.

[VBS08] S. Vlahinic, D. Brnobic, and N. Stojkovic, Indices for harmonic distortion monitoring of power distribution systems, *Instrumentation and Measurement Technology Conference Proceedings*, May 12–15, 2008, pages 421–425.

[Wat05] N.R. Watson, Advanced harmonic assessment, Power Engineering Society General Meeting, June 12–16, 2005, pages 2230–2235.

[Wat47] E.A. Watson et al., Liquid fuel pump governor. U.S. Patent 2,429,005 issued October 14, 1947, available online: http://www.uspto.gov (accessed February 25, 2013).

[Way-- ] Information on the Capstone microturbines in Waynesburg, Pennsylvania, available online: http://www.microturbine.com/_docs/CS_CAP387_Dominion_lowres.pdf (accessed April 11, 2013).

[WBH00] W.B. Hall et al., Communications processor remote host and multiple unit control devices and methods for micropower generation systems. U.S. Patent 6,055,163, issued April 25, 2000, available online: http://www.uspto.gov (accessed March 22, 2013).

[WCo74] W.C. Cornelius et al., Combustion method and apparatus. U.S. Patent 3,958,413, issued May 25, 1976, available online: http://www.uspto.gov (accessed March 10, 2013).

[WCP75] W.C. Pferfferle et al., Catalytically-supported thermal combustion. U.S. Patent 3,928,961, issued December 30, 1975, available online: http://www.uspto.gov (accessed March 17, 2013).

[WRo72] W. Rowen et al., Constant power control system for gas turbine. U.S. Patent 3,639,076, issued July 14, 1972, available online: http://www.uspto.gov (accessed February 28, 2013).

[WRR02] W.R. Ryan, Recuperator for use with turbine/turbo-alternator. U.S. Patent 6,438,936, issued August 27, 2002, available online: http://www.uspto.gov (accessed March 15, 2013).

[YCM94] Y.H. Yan, C.S. Chen, C.S. Moo, and C.T. Hsu, Harmonic analysis for industrial customers, *IEEE Transactions on Industry Applications*, 30(2), 462–468, March–April 1994.

[YHa97] Y. Honda, Gas turbine apparatus and method of operating same on gaseous fuel. U.S. Patent 5,609,016, issued March 11, 1997, available online: http://www.uspto.gov (accessed March 17, 2013).

[YKg05] Y. Kang, Annular recuperator design. U.S. Patent 6,951,110, issued October 4, 2005, available online: http://www.uspto.gov (accessed March 15, 2013).

[YT09] T. Yu, J. Tong, and K.W. Chan, Study on microturbine as a back-up power supply for power grid black-start, Power and Energy Society General Meeting, July 26–30, 2009, pages 1–6.

[YWS03] Z. Ye, T.C.Y. Wang, G. Sinha, and R. Zhang, Efficiency comparison for microturbine power conditioning systems, Power Electronics Specialist Conference, Vol. 4, June 5–19, 2003, pages 1551–1556.

[ZT08] D. Zhong, L.M. Tolbert, J.N. Chiasson, and B. Ozpineci, Reduced switching-frequency active harmonic elimination for multilevel converters, *IEEE Transactions on Industrial Electronics*, 55(4), 1761–1770, April 2008.

# Appendix 1

Sample Measurements for the Power Peak Trend of the TA-100

| Time (hours.minutes. seconds, centiseconds) | Phase A kW($P_{max}$) (kW) | Phase B kW($P_{max}$) (kW) | Phase C kW($P_{max}$) (kW) |
|---|---|---|---|
| 11.23.04,000 | –0.0427 | –0.0327 | 0.0338 |
| 11.23.06,000 | –0.0427 | –0.0327 | 0.0338 |
| 11.23.08,000 | –0.0427 | –0.0327 | 0.0328 |
| 11.23.10,000 | –0.0427 | –0.0327 | 0.0328 |
| 11.23.12,000 | –0.0427 | –0.0327 | 0.0328 |
| 11.23.14,000 | –0.0427 | –0.0327 | 0.0328 |
| 11.23.16,000 | –0.0427 | –0.0327 | 0.0328 |
| 11.23.18,000 | –0.0427 | –0.0327 | 0.0328 |
| 11.23.20,000 | –0.0427 | –0.0327 | 0.0328 |
| 11.23.22,000 | –0.0427 | –0.0327 | 0.0328 |
| 11.23.24,000 | –0.0427 | –0.0327 | 0.0328 |
| 11.23.26,000 | –0.0427 | –0.0327 | 0.0328 |
| 11.23.28,000 | –0.0427 | –0.0327 | 0.0328 |
| 11.23.30,000 | –0.0427 | –0.0327 | 0.0328 |
| 11.23.32,000 | –0.0427 | –0.0327 | 0.0328 |
| 11.23.34,000 | –0.0427 | –0.0327 | 0.0328 |
| 11.23.38,000 | –1.1251 | –0.5362 | 0.0347 |
| 11.23.40,000 | –0.6161 | –0.3177 | 0.0338 |
| 11.23.42,000 | –0.6341 | –0.3311 | 0.0338 |
| 11.23.44,000 | –0.6332 | –0.3302 | 0.0338 |
| 11.23.46,000 | –0.6313 | –0.3273 | 0.0338 |
| 11.23.48,000 | –0.6616 | –0.4842 | 0.0328 |
| 11.23.50,000 | –0.563 | –0.3908 | 0.0328 |
| 11.23.54,000 | –0.6625 | –0.3581 | –0.0174 |
| 11.23.56,000 | –0.6483 | –0.3437 | –0.0039 |
| 11.23.58,000 | –0.6379 | –0.3331 | 0.0068 |
| 11.24.00,000 | –0.6389 | –0.334 | 0.0029 |
| 11.24.02,000 | –0.6502 | –0.3456 | –0.0048 |
| 11.24.04,000 | –0.8445 | –0.4842 | –0.0309 |
| 11.24.06,000 | –0.8891 | –0.5198 | –0.056 |
| 11.24.08,000 | –0.9033 | –0.5371 | –0.0811 |
| 11.24.10,000 | –0.9317 | –0.566 | –0.1081 |
| 11.24.12,000 | –0.8114 | –0.4929 | –0.139 |
| 11.24.14,000 | –0.8009 | –0.4909 | –0.1651 |
| 11.24.16,000 | –0.8246 | –0.5198 | –0.195 |

*(Continued)*

| Time (hours.minutes. seconds, centiseconds) | Phase A kW($P_{max}$) (kW) | Phase B kW($P_{max}$) (kW) | Phase C kW($P_{max}$) (kW) |
|---|---|---|---|
| 11.24.20,000 | –0.8929 | –0.5785 | –0.2626 |
| 11.24.22,000 | –0.9327 | –0.619 | –0.3002 |
| 11.24.24,000 | –0.9697 | –0.6527 | –0.3417 |
| 11.24.26,000 | –1.0104 | –0.6969 | –0.3851 |
| 11.24.28,000 | –1.054 | –0.7374 | –0.4305 |
| 11.24.30,000 | –1.1024 | –0.7845 | –0.4797 |
| 11.24.32,000 | –1.1564 | –0.8375 | –0.528 |
| 11.24.34,000 | –1.2076 | –0.8895 | –0.5801 |
| 11.24.36,000 | –1.2644 | –0.9424 | –0.64 |
| 11.24.38,000 | –1.3346 | –1.0088 | –0.7046 |
| 11.24.40,000 | –1.3933 | –1.0666 | –0.7713 |
| 11.24.42,000 | –1.4625 | –1.1388 | –0.8398 |
| 11.24.46,000 | –1.6057 | –1.2783 | –0.9942 |
| 11.24.48,000 | –1.6862 | –1.3544 | –1.0772 |
| 11.24.50,000 | –1.7374 | –1.4073 | –1.1197 |
| 11.24.52,000 | –1.599 | –1.2697 | –0.9817 |
| 11.24.54,000 | –1.581 | –1.2591 | –0.9672 |
| 11.24.56,000 | –1.5858 | –1.2591 | –0.9643 |
| 11.24.58,000 | –1.5829 | –1.2591 | –0.9682 |
| 11.25.00,000 | –1.5801 | –1.2591 | –0.9653 |
| 11.25.02,000 | –1.5763 | –1.2629 | –0.9682 |
| 11.25.04,000 | –1.5763 | –1.262 | –0.9643 |
| 11.25.08,000 | –1.4986 | –1.1927 | –0.8871 |
| 11.25.10,000 | –0.837 | –0.5342 | –0.1979 |
| 11.25.12,000 | –0.673 | –0.3648 | –0.0145 |
| 11.25.14,000 | –0.6389 | –0.3311 | 0.0193 |
| 11.25.16,000 | –2.9782 | –2.6818 | –2.335 |
| 11.25.18,000 | –2.9924 | –2.7117 | –2.3717 |
| 11.25.20,000 | –3.2502 | –2.9639 | –2.6352 |
| 11.25.22,000 | –3.3061 | –3.0207 | –2.6931 |
| 11.25.24,000 | –3.3563 | –3.0804 | –2.753 |
| 11.25.26,000 | –3.4009 | –3.1275 | –2.8022 |
| 11.25.28,000 | –3.4701 | –3.1939 | –2.863 |
| 11.25.30,000 | –3.5251 | –3.245 | –2.918 |
| 11.25.32,000 | –3.5781 | –3.3046 | –2.974 |
| 11.25.36,000 | –3.6701 | –3.3894 | –3.0686 |
| 11.25.38,000 | –3.7099 | –3.4153 | –3.1062 |
| 11.25.40,000 | –3.7241 | –3.4375 | –3.1304 |
| 11.25.42,000 | –3.6767 | –3.3884 | –3.0754 |
| 11.25.44,000 | –3.8284 | –3.526 | –3.2259 |
| 11.25.46,000 | –3.9194 | –3.6223 | –3.3118 |
| 11.25.50,000 | –4.0833 | –3.7985 | –3.4895 |
| 11.25.52,000 | –4.1601 | –3.8697 | –3.5647 |

| Time (hours.minutes. seconds, centiseconds) | Phase A kW($P_{max}$) (kW) | Phase B kW($P_{max}$) (kW) | Phase C kW($P_{max}$) (kW) |
|---|---|---|---|
| 11.25.54,000 | –4.3061 | –4.0122 | –3.7124 |
| 11.25.56,000 | –4.3791 | –4.0911 | –3.8003 |
| 11.25.58,000 | –4.4758 | –4.1825 | –3.8804 |
| 11.26.00,000 | –4.2597 | –3.9717 | –3.6719 |
| 11.26.02,000 | –3.6056 | –3.3229 | –2.9991 |
| 11.26.04,000 | –3.1251 | –2.8426 | –2.5116 |
| 11.26.06,000 | –2.8663 | –2.5759 | –2.2327 |
| 11.26.08,000 | –2.8293 | –2.5374 | –2.1941 |
| 11.26.10,000 | –2.8312 | –2.5403 | –2.1999 |
| 11.26.12,000 | –2.8075 | –2.5134 | –2.1728 |
| 11.26.14,000 | –2.7914 | –2.497 | –2.1574 |
| 11.26.16,000 | –2.8692 | –2.5711 | –2.2307 |
| 11.26.18,000 | –2.9308 | –2.6337 | –2.2973 |
| 11.26.20,000 | –2.8237 | –2.5259 | –2.1844 |
| 11.26.22,000 | –2.8483 | –2.549 | –2.2076 |
| 11.26.26,000 | –2.8673 | –2.5711 | –2.2307 |
| 11.26.28,000 | –2.8739 | –2.5779 | –2.2385 |
| 11.26.32,000 | –2.8824 | –2.5798 | –2.2452 |
| 11.26.34,000 | –2.8824 | –2.5779 | –2.2423 |
| 11.26.36,000 | –2.8815 | –2.5808 | –2.2443 |
| 11.26.38,000 | –2.8739 | –2.574 | –2.2375 |
| 11.26.40,000 | –2.871 | –2.5721 | –2.2356 |
| 11.26.42,000 | –2.8616 | –2.5663 | –2.2269 |
| 11.26.44,000 | –2.8568 | –2.5567 | –2.2211 |
| 11.26.46,000 | –2.8473 | –2.549 | –2.2124 |
| 11.26.48,000 | –2.8455 | –2.5423 | –2.2047 |
| 11.26.50,000 | –2.836 | –2.5336 | –2.1979 |
| 11.26.52,000 | –2.8521 | –2.55 | –2.2182 |
| 11.26.54,000 | –2.9611 | –2.6684 | –2.2472 |
| 11.26.56,000 | –2.9924 | –2.6992 | –2.2433 |
| 11.26.58,000 | –2.9829 | –2.6867 | –2.2336 |
| 11.27.00,000 | –2.9744 | –2.677 | –2.224 |
| 11.27.02,000 | –2.963 | –2.6645 | –2.2066 |
| 11.27.04,000 | –2.9459 | –2.651 | –2.1902 |
| 11.27.06,000 | –2.9459 | –2.6491 | –2.1941 |
| 11.27.08,000 | –2.9497 | –2.6549 | –2.197 |
| 11.27.12,000 | –2.9734 | –2.6857 | –2.2182 |
| 11.27.16,000 | –2.908 | –2.626 | –2.1834 |
| 11.27.18,000 | –2.8928 | –2.6135 | –2.1748 |
| 11.27.20,000 | –2.8891 | –2.6068 | –2.1738 |
| 11.27.22,000 | –2.8853 | –2.6039 | –2.1719 |
| 11.27.24,000 | –2.8862 | –2.6039 | –2.1767 |

(*Continued*)

| Time (hours.minutes. seconds, centiseconds) | Phase A kW($P_{max}$) (kW) | Phase B kW($P_{max}$) (kW) | Phase C kW($P_{max}$) (kW) |
|---|---|---|---|
| 11.27.26,000 | –2.8862 | –2.6019 | –2.1786 |
| 11.27.28,000 | –2.8625 | –2.5817 | –2.1661 |
| 11.27.30,000 | –2.8369 | –2.5577 | –2.14 |
| 11.27.32,000 | –2.8189 | –2.5394 | –2.1246 |
| 11.27.34,000 | –2.8237 | –2.5432 | –2.1313 |
| 11.27.36,000 | –2.8341 | –2.5528 | –2.1468 |
| 11.27.38,000 | –2.836 | –2.5567 | –2.1535 |
| 11.27.40,000 | –2.8341 | –2.5577 | –2.1555 |
| 11.27.42,000 | –2.8189 | –2.5432 | –2.1468 |
| 11.27.44,000 | –2.8113 | –2.5336 | –2.139 |
| 11.27.46,000 | –2.8 | –2.522 | –2.1371 |
| 11.27.48,000 | –2.7943 | –2.522 | –2.1323 |
| 11.27.50,000 | –2.7838 | –2.5134 | –2.1275 |
| 11.27.54,000 | –2.781 | –2.5057 | –2.1226 |
| 11.27.56,000 | –2.7753 | –2.5018 | –2.1226 |
| 11.27.58,000 | –2.7725 | –2.5028 | –2.1246 |
| 11.28.00,000 | –2.7687 | –2.4989 | –2.1226 |
| 11.28.02,000 | –2.7763 | –2.5047 | –2.1255 |
| 11.28.06,000 | –2.7696 | –2.5038 | –2.1313 |
| 11.28.08,000 | –2.7658 | –2.4999 | –2.1265 |
| 11.28.10,000 | –2.7649 | –2.5009 | –2.1313 |
| 11.28.12,000 | –2.7639 | –2.4999 | –2.1255 |
| 11.28.14,000 | –2.762 | –2.4999 | –2.1284 |
| 11.28.16,000 | –2.762 | –2.4961 | –2.1265 |
| 11.28.18,000 | –2.7365 | –2.4749 | –2.0985 |
| 11.28.20,000 | –2.7184 | –2.4547 | –2.0821 |
| 11.28.22,000 | –2.7241 | –2.4595 | –2.0975 |
| 11.28.24,000 | –2.7232 | –2.4595 | –2.0956 |
| 11.28.26,000 | –2.7279 | –2.4624 | –2.1014 |
| 11.28.28,000 | –2.7289 | –2.4633 | –2.1033 |
| 11.28.30,000 | –2.727 | –2.4614 | –2.1072 |
| 11.28.32,000 | –2.7327 | –2.4633 | –2.1082 |
| 11.28.34,000 | –2.7317 | –2.4662 | –2.112 |
| 11.28.36,000 | –2.7336 | –2.471 | –2.1139 |
| 11.28.38,000 | –2.7327 | –2.471 | –2.1139 |
| 11.28.40,000 | –2.7412 | –2.4739 | –2.1159 |
| 11.28.42,000 | –2.7014 | –2.4335 | –2.0792 |
| 11.28.44,000 | –2.6928 | –2.4287 | –2.0763 |
| 11.28.46,000 | –2.709 | –2.4441 | –2.0898 |
| 11.28.48,000 | –2.7175 | –2.4508 | –2.0956 |
| 11.28.50,000 | –2.7232 | –2.4585 | –2.1004 |
| 11.28.52,000 | –2.7317 | –2.4672 | –2.1091 |
| 11.28.56,000 | –2.7147 | –2.4489 | –2.0937 |

| Time (hours.minutes. seconds, centiseconds) | Phase A kW($P_{max}$) (kW) | Phase B kW($P_{max}$) (kW) | Phase C kW($P_{max}$) (kW) |
|---|---|---|---|
| 11.28.58,000 | –2.7061 | –2.4402 | –2.0879 |
| 11.29.00,000 | –2.6985 | –2.4373 | –2.0821 |
| 11.29.02,000 | –2.7042 | –2.4383 | –2.084 |
| 11.29.04,000 | –2.7014 | –2.4373 | –2.0811 |
| 11.29.06,000 | –2.6957 | –2.4325 | –2.0773 |
| 11.29.08,000 | –2.6938 | –2.4325 | –2.0782 |
| 11.29.10,000 | –2.691 | –2.4306 | –2.0715 |
| 11.29.12,000 | –2.6947 | –2.4296 | –2.0753 |
| 11.29.16,000 | –2.6928 | –2.4335 | –2.0802 |
| 11.29.18,000 | –2.6928 | –2.4325 | –2.0802 |
| 11.29.20,000 | –2.6938 | –2.4335 | –2.0802 |
| 11.29.22,000 | –2.6966 | –2.4373 | –2.0831 |
| 11.29.24,000 | –2.6938 | –2.4344 | –2.085 |
| 11.29.26,000 | –2.6947 | –2.4393 | –2.0831 |
| 11.29.28,000 | –2.6976 | –2.4412 | –2.086 |
| 11.29.30,000 | –2.6938 | –2.4335 | –2.0831 |
| 11.29.32,000 | –2.6995 | –2.4364 | –2.084 |
| 11.29.34,000 | –2.709 | –2.447 | –2.0898 |
| 11.29.36,000 | –2.6995 | –2.4412 | –2.0917 |
| 11.29.38,000 | –2.7042 | –2.4412 | –2.0888 |
| 11.29.40,000 | –2.7033 | –2.4421 | –2.0917 |
| 11.29.42,000 | –2.7071 | –2.445 | –2.0927 |
| 11.29.46,000 | –2.6891 | –2.4287 | –2.0773 |
| 11.29.48,000 | –2.6872 | –2.4239 | –2.0715 |
| 11.29.50,000 | –2.6834 | –2.419 | –2.0666 |
| 11.29.52,000 | –2.6843 | –2.419 | –2.0657 |
| 11.29.54,000 | –2.6815 | –2.4142 | –2.0666 |
| 11.29.58,000 | –2.6796 | –2.4094 | –2.0618 |
| 11.30.00,000 | –2.6805 | –2.4123 | –2.0599 |
| 11.30.02,000 | –2.6805 | –2.4142 | –2.0589 |
| 11.30.04,000 | –2.6777 | –2.4075 | –2.0589 |
| 11.30.06,000 | –2.6777 | –2.4056 | –2.056 |
| 11.30.08,000 | –2.6767 | –2.4056 | –2.057 |
| 11.30.10,000 | –2.6767 | –2.4046 | –2.0531 |
| 11.30.12,000 | –2.6796 | –2.4104 | –2.057 |
| 11.30.14,000 | –2.6786 | –2.4113 | –2.0551 |
| 11.30.16,000 | –2.6786 | –2.4142 | –2.058 |
| 11.30.18,000 | –2.6786 | –2.4123 | –2.058 |
| 11.30.20,000 | –2.6815 | –2.4142 | –2.0609 |
| 11.30.22,000 | –2.6796 | –2.4104 | –2.0609 |
| 11.30.24,000 | –2.6767 | –2.4085 | –2.0638 |
| 11.30.26,000 | –2.6805 | –2.4113 | –2.0609 |

*(Continued)*

| Time (hours.minutes. seconds, centiseconds) | Phase A kW($P_{max}$) (kW) | Phase B kW($P_{max}$) (kW) | Phase C kW($P_{max}$) (kW) |
|---|---|---|---|
| 11.30.28,000 | −2.6786 | −2.4133 | −2.0599 |
| 11.30.30,000 | −2.6739 | −2.4104 | −2.0551 |
| 11.30.32,000 | −2.6739 | −2.4094 | −2.058 |
| 11.30.36,000 | −2.3715 | −2.136 | −1.7742 |
| 11.30.40,000 | −1.9213 | −1.6942 | −1.334 |
| 11.30.42,000 | −1.6995 | −1.467 | −1.1159 |
| 11.30.44,000 | −1.4777 | −1.2475 | −0.8842 |
| 11.30.46,000 | −1.2682 | −1.0329 | −0.6844 |
| 11.30.48,000 | −1.0578 | −0.8269 | −0.4614 |
| 11.30.50,000 | −0.837 | −0.6122 | −0.2558 |
| 11.30.52,000 | −0.5175 | −0.2743 | −0.0319 |
| 11.30.54,000 | −0.2863 | −0.0703 | 0.1776 |
| 11.30.56,000 | −0.0758 | 0.1675 | 0.4016 |
| 11.30.58,000 | 0.1299 | 0.3716 | 0.6274 |
| 11.31.00,000 | 0.3706 | 0.6093 | 0.8398 |
| 11.31.02,000 | 0.581 | 0.8259 | 1.0763 |
| 11.31.04,000 | 0.8152 | 1.0589 | 1.2838 |
| 11.31.06,000 | 1.0104 | 1.2629 | 1.5078 |
| 11.31.08,000 | 1.2407 | 1.4747 | 1.7105 |
| 11.31.10,000 | 1.436 | 1.6942 | 1.9286 |
| 11.31.12,000 | 1.6616 | 1.8992 | 2.1526 |
| 11.31.14,000 | 1.8739 | 2.1408 | 2.363 |
| 11.31.16,000 | 2.0948 | 2.3343 | 2.5927 |
| 11.31.20,000 | 2.527 | 2.7906 | 3.0377 |
| 11.31.22,000 | 2.7649 | 3.0168 | 3.2366 |
| 11.31.26,000 | 3.218 | 3.4596 | 3.6883 |
| 11.31.28,000 | 3.4161 | 3.7022 | 3.919 |
| 11.31.30,000 | 3.6587 | 3.9101 | 4.1487 |
| 11.31.32,000 | 3.8691 | 4.1556 | 4.3582 |
| 11.31.34,000 | 4.0995 | 4.3606 | 4.6092 |
| 11.31.36,000 | 4.3241 | 4.609 | 4.8022 |
| 11.31.38,000 | 4.5345 | 4.8198 | 5.06 |
| 11.31.40,000 | 4.78 | 5.0556 | 5.2607 |
| 11.31.42,000 | 4.9686 | 5.2828 | 5.5175 |
| 11.31.44,000 | 5.2255 | 5.4946 | 5.7231 |
| 11.31.46,000 | 5.4122 | 5.7314 | 5.9528 |
| 11.31.48,000 | 5.6653 | 5.9307 | 6.1835 |
| 11.31.50,000 | 5.853 | 6.1954 | 6.393 |
| 11.31.52,000 | 6.0937 | 6.3812 | 6.6382 |
| 11.31.54,000 | 6.2956 | 6.6334 | 6.8341 |
| 11.31.56,000 | 6.525 | 6.823 | 7.0832 |
| 11.31.58,000 | 6.7392 | 7.0829 | 7.2839 |
| 11.32.02,000 | 7.1876 | 7.5305 | 7.7183 |

| Time (hours.minutes. seconds, centiseconds) | Phase A kW($P_{max}$) (kW) | Phase B kW($P_{max}$) (kW) | Phase C kW($P_{max}$) (kW) |
|---|---|---|---|
| 11.32.04,000 | 7.4151 | 7.7201 | 7.9722 |
| 11.32.06,000 | 7.6188 | 7.9839 | 8.1797 |
| 11.32.08,000 | 7.8586 | 8.1639 | 8.4162 |
| 11.32.10,000 | 8.0615 | 8.4354 | 8.6151 |
| 11.32.12,000 | 8.2966 | 8.6192 | 8.8573 |
| 11.32.16,000 | 8.7496 | 9.0678 | 9.3052 |
| 11.32.18,000 | 8.942 | 9.3344 | 9.535 |

# Appendix 2

Voltages and Currents Harmonics in the Starting Moment of the Microturbine

| Harmonic Order | CHA Volts | CHB Volts | CHC Volts | CHA Amps | CHB Amps | CHC Amps |
|---|---|---|---|---|---|---|
| H02 | 0 | 0.01 | 0.02 | 0.00126 | 0.00114 | 0.00141 |
| H03 | 1 | 0.55 | 1.03 | 0.32261 | 0.32418 | 0.00414 |
| H04 | 0.02 | 0.02 | 0.03 | 0.0015 | 0.00122 | 0.00023 |
| H05 | 4.95 | 4.73 | 4.63 | 0.15055 | 0.30065 | 0.15698 |
| H06 | 0.02 | 0.02 | 0 | 0.00158 | 0.00228 | 0.00383 |
| H07 | 2.1 | 1.84 | 2.02 | 0.27605 | 0.19034 | 0.09059 |
| H08 | 0.01 | 0.02 | 0.01 | 0.00165 | 0.00155 | 0.00258 |
| H09 | 0.58 | 0.84 | 0.67 | 0.13172 | 0.13588 | 0.00476 |
| H10 | 0.02 | 0.01 | 0 | 0.00142 | 0.00049 | 0.00102 |
| H11 | 1.26 | 1.24 | 1.15 | 0.1099 | 0.13946 | 0.07466 |
| H12 | 0.01 | 0.02 | 0.03 | 0.00118 | 0.00261 | 0.0018 |
| H13 | 0.46 | 0.4 | 0.63 | 0.09115 | 0.06065 | 0.04881 |
| H14 | 0 | 0.01 | 0.01 | 0.00213 | 0.00179 | 0.00086 |
| H15 | 0.06 | 0.2 | 0.34 | 0.05026 | 0.03737 | 0.01804 |
| H16 | 0 | 0.01 | 0 | 0.0015 | 0.00081 | 0.00195 |
| H17 | 0.14 | 0.08 | 0.01 | 0.04782 | 0.04095 | 0.00648 |
| H18 | 0.03 | 0.04 | 0.03 | 0.00299 | 0.00179 | 0.00164 |
| H19 | 0.11 | 0.18 | 0.1 | 0.03135 | 0.04201 | 0.01851 |
| H20 | 0.01 | 0 | 0.01 | 0.00071 | 0.00057 | 0.00203 |
| H21 | 0.13 | 0.13 | 0.16 | 0.01993 | 0.01286 | 0.00594 |
| H22 | 0 | 0.01 | 0.01 | 0.00063 | 0.00155 | 0.00062 |
| H23 | 0.03 | 0.04 | 0.04 | 0.01379 | 0.01579 | 0.00523 |
| H24 | 0.04 | 0.06 | 0.06 | 0.00181 | 0.00114 | 0.00133 |
| H25 | 0.04 | 0.07 | 0.04 | 0.01568 | 0.021 | 0.00601 |
| H26 | 0 | 0.01 | 0.02 | 0.00213 | 0.00122 | 0.00055 |
| H27 | 0.04 | 0.07 | 0.08 | 0.01481 | 0.01221 | 0.00164 |
| H28 | 0.01 | 0.01 | 0.02 | 0.00323 | 0.00374 | 0.00375 |
| H29 | 0.04 | 0.03 | 0.01 | 0.01134 | 0.0127 | 0.0043 |
| H30 | 0.07 | 0.09 | 0.09 | 0.0026 | 0.00155 | 0.00109 |
| H31 | 0.05 | 0.02 | 0.03 | 0.01442 | 0.01058 | 0.00734 |
| H32 | 0.01 | 0.02 | 0.01 | 0.00252 | 0.00236 | 0.00266 |
| H33 | 0.07 | 0.09 | 0.09 | 0.00906 | 0.00969 | 0.00211 |
| H34 | 0.01 | 0.02 | 0.03 | 0.00662 | 0.00749 | 0.0064 |
| H35 | 0 | 0.02 | 0.03 | 0.00654 | 0.00961 | 0.00586 |
| H36 | 0.12 | 0.15 | 0.14 | 0.00252 | 0.00415 | 0.00125 |

*(Continued)*

| Harmonic Order | CHA Volts | CHB Volts | CHC Volts | CHA Amps | CHB Amps | CHC Amps |
|---|---|---|---|---|---|---|
| H37 | 0.02 | 0.01 | 0.04 | 0.00961 | 0.01099 | 0.00703 |
| H38 | 0.02 | 0.04 | 0.02 | 0.00733 | 0.00521 | 0.00758 |
| H39 | 0.1 | 0.11 | 0.12 | 0.00693 | 0.00277 | 0.0018 |
| H40 | 0.03 | 0.03 | 0.05 | 0.01134 | 0.00904 | 0.01023 |
| H41 | 0.01 | 0.01 | 0.01 | 0.00228 | 0.00497 | 0.00437 |
| H42 | 0.19 | 0.22 | 0.21 | 0.00315 | 0.00513 | 0.00219 |
| H43 | 0.02 | 0.03 | 0.03 | 0.00867 | 0.00741 | 0.00289 |
| H44 | 0.03 | 0.03 | 0.02 | 0.01008 | 0.00635 | 0.01007 |
| H45 | 0.06 | 0.06 | 0.06 | 0.00197 | 0.00659 | 0.00531 |
| H46 | 0.04 | 0.01 | 0.05 | 0.01095 | 0.00993 | 0.01414 |
| H47 | 0.02 | 0.02 | 0.01 | 0.00575 | 0.00643 | 0.00617 |
| H48 | 0.17 | 0.19 | 0.18 | 0.00126 | 0.00586 | 0.00211 |
| H49 | 0.01 | 0.01 | 0.01 | 0.00512 | 0.00383 | 0.00266 |
| H50 | 0.02 | 0 | 0 | 0.00575 | 0.00138 | 0.00289 |

# Appendix 3

Voltages and Currents Harmonics 10 Minutes after the Microturbine Start

| Harmonic Order | CHA Volts | CHB Volts | CHC Volts | CHA Amps | CHB Amps | CHC Amps |
|---|---|---|---|---|---|---|
| H02 | 0.02 | 0.02 | 0.03 | 0.042 | 0.049 | 0.091 |
| H03 | 1.12 | 0.58 | 1.05 | 0.425 | 0.623 | 0.219 |
| H04 | 0.02 | 0.03 | 0.03 | 0.036 | 0.006 | 0.037 |
| H05 | 5.03 | 4.78 | 4.63 | 0.892 | 0.965 | 0.731 |
| H06 | 0.02 | 0.01 | 0.01 | 0.03 | 0.037 | 0.061 |
| H07 | 1.89 | 1.69 | 1.83 | 0.844 | 0.824 | 0.768 |
| H08 | 0.01 | 0.02 | 0.02 | 0.03 | 0.031 | 0.061 |
| H09 | 0.65 | 0.9 | 0.74 | 0.096 | 0.11 | 0.037 |
| H10 | 0.02 | 0.03 | 0.01 | 0.012 | 0.024 | 0.018 |
| H11 | 1.11 | 1.06 | 1.04 | 0.377 | 0.324 | 0.262 |
| H12 | 0 | 0.03 | 0.03 | 0 | 0.006 | 0 |
| H13 | 0.41 | 0.46 | 0.65 | 0.132 | 0.159 | 0.171 |
| H14 | 0 | 0 | 0 | 0.012 | 0.012 | 0.006 |
| H15 | 0.18 | 0.3 | 0.43 | 0.06 | 0.055 | 0.055 |
| H16 | 0 | 0.01 | 0 | 0.012 | 0.012 | 0.018 |
| H17 | 0.09 | 0.08 | 0.03 | 0.06 | 0.085 | 0.067 |
| H18 | 0.01 | 0.02 | 0.02 | 0.036 | 0.012 | 0.037 |
| H19 | 0.12 | 0.19 | 0.13 | 0.06 | 0.092 | 0.079 |
| H20 | 0.01 | 0.01 | 0.01 | 0.042 | 0.037 | 0.03 |
| H21 | 0.08 | 0.07 | 0.01 | 0.03 | 0.006 | 0.03 |
| H22 | 0.02 | 0.01 | 0 | 0.018 | 0.031 | 0.049 |
| H23 | 0.07 | 0.06 | 0.11 | 0.108 | 0.104 | 0.061 |
| H24 | 0.03 | 0.05 | 0.01 | 0.018 | 0.073 | 0.073 |
| H25 | 0.06 | 0.09 | 0.07 | 0.072 | 0.092 | 0.03 |
| H26 | 0.01 | 0.01 | 0.01 | 0.03 | 0.049 | 0.037 |
| H27 | 0.16 | 0.14 | 0.14 | 0.006 | 0.024 | 0.018 |
| H28 | 0.01 | 0.01 | 0 | 0.024 | 0.006 | 0.006 |
| H29 | 0.04 | 0.06 | 0.06 | 0.048 | 0.043 | 0.043 |
| H30 | 0.05 | 0.07 | 0.07 | 0.006 | 0.018 | 0.018 |
| H31 | 0.01 | 0 | 0.01 | 0.036 | 0.043 | 0.043 |
| H32 | 0.02 | 0.01 | 0.01 | 0.042 | 0.018 | 0.037 |
| H33 | 0.17 | 0.2 | 0.19 | 0.006 | 0.012 | 0.006 |
| H34 | 0.01 | 0 | 0.02 | 0 | 0.006 | 0.006 |
| H35 | 0.03 | 0.04 | 0.03 | 0.012 | 0.012 | 0.018 |
| H36 | 0.08 | 0.11 | 0.09 | 0.012 | 0.024 | 0.018 |

*(Continued)*

| Harmonic Order | CHA Volts | CHB Volts | CHC Volts | CHA Amps | CHB Amps | CHC Amps |
|---|---|---|---|---|---|---|
| H37 | 0.05 | 0 | 0.04 | 0.018 | 0.037 | 0.012 |
| H38 | 0.01 | 0.01 | 0.01 | 0 | 0.012 | 0.018 |
| H39 | 0.17 | 0.18 | 0.19 | 0.024 | 0.031 | 0.006 |
| H40 | 0.01 | 0 | 0.02 | 0.006 | 0.012 | 0.006 |
| H41 | 0.02 | 0.02 | 0.01 | 0.036 | 0.031 | 0.03 |
| H42 | 0.05 | 0.07 | 0.07 | 0.018 | 0.012 | 0.006 |
| H43 | 0.04 | 0.02 | 0.05 | 0.024 | 0.024 | 0.037 |
| H44 | 0.03 | 0.01 | 0.01 | 0.006 | 0.018 | 0.012 |
| H45 | 0.05 | 0.08 | 0.07 | 0.006 | 0.006 | 0 |
| H46 | 0.03 | 0.03 | 0.01 | 0.012 | 0.006 | 0 |
| H47 | 0.02 | 0.01 | 0 | 0.012 | 0.012 | 0.018 |
| H48 | 0.03 | 0.07 | 0.06 | 0.012 | 0.031 | 0 |
| H49 | 0.02 | 0.03 | 0.04 | 0.036 | 0.024 | 0.03 |
| H50 | 0.02 | 0.03 | 0.01 | 0.024 | 0.024 | 0.006 |

# Index

## C

## Q